Being = Space × Action

Being = Space × Action

Searches for Freedom of Mind
through Mathematics, Art, and Mysticism

Io #41

edited by
Charles Stein

North Atlantic Books
Berkeley, California

Acknowledgement

I am very grateful to Professor A.S. Yessenin-Volpin for permitting *Io* to publish this translation of *On The Logic of The Moral Sciences*.

Being = Space × Action
Searches for Freedom of Mind through Mathematics,
Art, and Mysticism

Copyright © 1988 by Charles Stein
Copyrights of individual pieces to their respective authors

ISBN 1-55643-043-4 (paperback)
ISBN 1-55643-044-2 (cloth)

Published by North Atlantic Books
2800 Woolsey Street
Berkeley, California 94705

This is issue #41 in the *Io* series. Subscribers may purchase four consecutive issues beginning either with the current one or #42 for $32 (United States) or $40 (elsewhere).

Cover and book design by Paula Morrison
Cover art: Abstract Tactical Configuration, by Christer Hennix
Typeset by Campaigne and Somit Typography

This project is partially supported by the National Endowment for the Arts, a Federal agency, Washington, D.C.

Being = Space × Action: Searches for Freedom of Mind through Mathematics, Art, and Mysticism is sponsored by the Society for the Study of Native Arts and Sciences, a nonprofit educational corporation whose goals are to develop an ecological and crosscultural perspective linking various scientific, social, and artistic fields; to nurture a holistic view of arts, sciences, humanities, and healing; and to publish and distribute literature on the relationship of mind, body, and nature.

Library of Congress Cataloging-in-Publication Data

Being = space × action: searches for freedom of mind through mathematics,
   art, and mysticism / edited by Charles Stein.
     p. cm.—(Io; no. 41)
    ISBN 1-55643-044-2: $30.00. — ISBN 1-55643-043-4 (pbk.): $16.95
   1. Aesthetics. 2. Mathematics. 3. Mysticism. I. Stein, Charles,
  1944- . II. Title: Being equals space times action. III. Series.
BH39.B3828   1989
111'.85—dc19
                                             89-2907
                                                CIP

Contents

Charles Stein
Introduction . . . 1

Don Byrd
Symbolic Nature . . . 55

A.S. Yessenin-Volpin
On the Logic of the Moral Sciences . . . 107

Henry Flynt and Christer Hennix
Philosophy of Concept Art . . . 155

Henry Flynt
Concept Art . . . 183

Henry Flynt
The Apprehension of Plurality . . . 191

George Quasha
Poems . . . 229

Charles Stein
Poems from *The Sad World* . . . 257

Christer Hennix
The Yellow Book . . . 271

Charles Stein
Notes Toward a Translation of Parmenides . . . 423

Introduction

Charles Stein

ༀ༔   །ཨ་བགས་པ་འཇམ་དཔལ་གཞོན་ནུར་གྱུར་པ་ལ་ཕྱག་འཚལ་ལོ།

It falls to me to introduce the various texts assembled in this issue of *Io*. I have decided that the most expeditious way of suggesting the thread of connectivity that runs through them is to begin with a somewhat personal narration of the circumstances which, in fact, have brought them together.

In the summer of 1982 I participated in a seminar at The Massachusetts Institute of Technology convened by the linguist Haj Ross. Ross gathered twelve artists from various disciplines to discuss what he took to be certain fundamental formal structures underlying the aesthetics not only of art, but of all creativity including mathematics and to some degree science. Occasionally attending the seminar were linguists, physicists and mathematicians; and, during the course of the summer, connections between mathematical structures, aesthetic notions of form, and eastern meditative practices were frequently commented upon. In my case, an interest in mathematical studies that had lain dormant since high school was awakened and I began to entertain the possibility of exploring the relationships between mathematical abstraction, artistic abstraction, and the abstract structures implicit in such practices as Zen meditation and other contemplative disciplines.

Two things particularly struck me at MIT: first, I became aware that certain mathematicians had, in fact, explicitly associated the states of mind involved in mathematical abstraction with contemplative concerns; second, that though most traditional contemplative practices encourage a *cessation* of cognitive activity, the very structures of the practices leading to such cessation are frequently themselves expressed or at least expressible in eminently abstract terms. One performs logical operations upon the continuum of awareness in order to curtail the proliferation and differentiation of events within it. Furthermore, upon developing their powers of concentration and continuous attention to a sufficient degree, meditators discover that cognitive operations are involved in the most concrete and least discursive aspects of awareness: spontaneous acts of identification and differentiation occur unremittingly and the imputation of existence to objects seems to subtend all conscious life. But such acts of identification, differentiation and imputation form the basis for mathematical abstraction as well.

When I returned from MIT I began informing myself about various branches of modern mathematical thought and philosophy with a view primarily to exploring these interconnections. By the middle of the fall of 1982 I had been introduced to Christer Hennix, and something of an apprenticeship in the intricacies of his mathematical work began. Hennix is a Swedish-born musician, "concept" artist and poet, as well as the holder of an advanced degree in philosophy. Since 1980 Hennix has held various positions as a visiting professor of logic at MIT AI laboratory, Cambridge, Mass., and as an assistant professor of mathematics and computer science at State University of New York at New Platz, N.Y. He is the principal student of the founder of the Ultra-Intuitionist School of the Foundations of Mathematics, A.S. Yessenin-Volpin.

Inspired by this apprenticeship, in the winter of 1984 I succeeded in interesting two of my poet-associates, Don Byrd and George Quasha, in joining with me in a more or less formal series of seminars to center around Hennix's work, but to involve—in addition to mathematics and logic—poetic and other artistic practices, various facets of contemporary philosophy, Pre-Socratic thought and the Greek Mysteries, visionary poetics and experience, oriental, psychedelic and other forms of contemplative explorations. The seminars ran through the summer of 1984 under the rubric of "The Rhinebeck Institute." (Hennix lives in Rhinebeck New York and most of the sessions were held at his studio there.) Some of the poems by George Quasha included in this issue were read and discussed by us at some of the seminar sessions. The subject matter of Don Byrd's essay (which is a chapter from his current prose project, *The Poetics of Common Knowledge*), namely, the relationship between Cartesian mathematics and modern poetics, formed an

Introduction

important area of discussion.

In addition to the above-mentioned participants, the seminar was occasionally attended by Susan Quasha, Dawn Felicioni, psychologist Anne McClellan, composer Niel Rolnick, and curator (of the Blum Gallery at Bard College) Linda Weintraub.

Though never attending the seminar, Henry Flynt and A.S. Yessenin-Volpin were each in different ways important "presences." Most of us were studying their writings and their work was frequently alluded to during the sessions.

The remainder of this introduction will attempt to summarize a fraction of the material we discussed during those seminars under the following topics: the Parmenides Poem (a founding text of Western rational philosophy that also happens to be a visionary journey); the work of L.E.J. Brouwer (the Dutch mathematician and mystic who founded the Intuitionist School of foundations of mathematics); the work of A.S. Yessenin-Volpin (founder of Ultra-Intuitionism); and the work of Christer Hennix. I will leave the texts of Flynt, Quasha, Byrd —as well as my own poems—to speak for themselves.

The Parmenides Poem

A consistent concern throughout the seminar was the study, interpretation and translation of Parmenides. The seminar culminated with a public lecture given by Hennix on "Eleatic Set Theory," which I introduced with an earlier version of the translation printed at the end of this issue.

Parmenides is usually recognized as the beginning of the "rational" tradition in Western Philosophy. He is the first to leave a text behind in which the assertions are supported by definite arguments, and where belief demands proof.

Yet the Parmenides that appears in his fragmentary poem is not simply a rationalist in any familiar modern sense. He writes, for one thing, in verse, not prose. The occasion for his setting forth his arguments is a visionary journey; his mouth piece, a goddess; and the object of his vision, an enigmatic and mysterious totality that, though unnamed, is referred to by the locution "esti" (it is) or "on" (Being). The Goddess who speaks of Being is Dike or Justice, and thus though the technical philosophy is thoroughly abstract in content, its domain seems to be the whole of human action, and its attitude, ethical.

The Parmenides poem in fact opens a rather impressively large range of philosophical issues which remain unresolved to this day: the questions of the unity of knowledge, the unity of being, the unity of knowledge *and* being, the nature of proof, the nature of thought itself, the nature of justice, the relations between thought, proof, knowledge and justice, the possibility of rational cosmology and the relationship between logic and cosmology, the existence and nature of abstract entities, the privileged status of philosophical knowledge—all these questions *became* explicitly formulatable questions immediately following the production of Parmenides' poem, none of them are definitely resolved today, and the future of science, culture and the human spirit all depend to a far from trivial degree on how these questions come to be reflected upon.

But Parmenides was a visionary and a poet as well as a philosopher and mathematician, and as we were a group of poets concerned with philosophy, mathematics and visionary matters, we took the task of discussing, interpreting and translating Parmenides as an appropriate collective concern.

The text of the poem comes to us in fragments recorded by later philosophers: Aristotle and Simplicius, principally. It is generally assumed that it appeared in two parts: one delineating a "true" path, the other a representation of a path of mere opinion. The reason for this division is not clear from a study of the existing fragments. Numerous analyses have been suggested. It is also uncertain what proportion of the entire poem we are in possession of, and, in the case of certain passages, to which section they belong.

The poem is written in dactylic hexameters, the meter of the Homeric Poems and the Delphic oracular responses; however, the diction is not derived from the Homeric dialect and if the formulaic theory of the technique of Homeric poetry is correct, the Parmenidean poem does not employ it. Exactly why Parmenides wrote in hexameters when there was already precedent for philosophers to write without them is conjecture.

Though the thought of the poem is supported by argumentation, this thought manifests in the context of an ecstatic vision. Parmenides is taken on a journey in a chariot, and with the assistance of the daughters of the sun god, he ascends into a super-celestial region normally forbidden to mortal inspection. Parmenides' personal qualifications for the journey combine with a certain good fortune to create the conditions for his welcome by the goddess.

Parmenides' chariot journey resonates with similar voyages occurring in the texts of cultures geographically proximate to and contemporary with the Classical Greeks, cultures whose interrelation with the Hellenic world is speculative but by no means impossible. I am thinking, in particular, of the Vision of the Chariot in *Ezekiel* and the metaphor of the chariot in the *Upanishads*. A variant of the image of the chariot as a metaphor for a spiritual journey is taken up, of course, in Plato's *Phaedrus*.

Parmenides tells us that the horses that drive his chariot have the capacity to take him as far as he wishes to travel. I read the horses as images for his own mental capacities, the chariot as the vehicle of thought *cum* contemplation upon which he journeys. Parmenides has mastered his own intelligence and his own mind as far as that is possible, and this qualifies him for contact with a transcendent source of wisdom, the goddess herself.

The goddess herself is unique to Parmenides. I speculate that the reverence paid to the goddess justifies the use of Heroic verse. In any case the hexameters do not appear to be wielded ironically: Parmenides is not mocking the poetic tradition by using its style. Though his thought is a new vision of reality and a radical critique of ancestral myth and recent philosophy alike, something of the sense of the sacred as manifested in the poetic tradition is being honored and carried forward here. Again, though Parmenides radically decomposes previous thought, the sacred awe of the truth that such predecessor ontologies radiated in their poetries is being transmitted in a new mode.

The symbolism of the poem is a species of mystical allegory wherein persons and objects, though figurative, cannot be replaced by the abstractions for which in fact they stand. Dike is more than abstract "Justice," yet she is also more than an anthropomorphic goddess: Parmenides' conversation with her and the instruction he receives from her constitute a communication and instruction from Justice itself; and Justice is taken as somehow identical with the transcendent object, the It Is, of which the goddess speaks. That Dike speaks in arguments and proofs bespeaks the nature of Justice: Justice is inseparable from correct reasoning, and the rationality of Being conjoins its own theodicy.

The restrictions on reasoning that Justice demands are stringent. In particular, a certain form of negative reasoning or negative existence is radically forbidden. That which Is Not cannot be asserted To Be, and thus no creature made up partly of Being and partly of Non-Being can be granted any share in existence at all. Being is whole, self-identical, bounded, indivisible, and eternal. It is like a "well-rounded sphere."

It has often been wondered how Parmenides could have permitted himself numerous negative expressions in the delineation of Being, if the existence of non-being is to be so severely interdicted. But a recent study by Scot Austin (*Parmenides, Being, Bounds and Logic*, 1986, New Haven, Yale University Press) demonstrates that the ban on negative language is itself limited to a ban on the positive assertion of negative subjects: such constructions as "the non-existence of X exists." It does not disallow such assertions as "X does not have property y." Austin argues that Parmenides' language is consistently disciplined in this way, so that the charge of self-contradiction or verbal unconsciousness will not hold.

The image of the "well-rounded" sphere is extraordinary. It has often been taken literally, as if Parmenides meant to project a spherical cosmology of a material sort. In fact, the image is clearly a simile—it is meant to express the concept that mensurable distances, (such as the distances measured by geometers) have no application within Being. There is no distance as far as things of the mind are concerned. The aspect of the sphere that is intended in the simile is that which is contained in its formal definition: that all points on its surface are equidistant from its center. Similarly, all "points" or thoughts within Being are correlated without subdivision or hierarchical organization of any kind.

Parmenides' "sphere of being" is often, in recent literature, taken as the prototype of the kind of systematizing, totalizing cosmologies and philosophies which, imposing themselves on the world, justify hierarchically-organized social structures, mind-stunning super-systems, and all the political, social and theological tyrannies of western history. But it seems to me it is important to notice that whatever the aspirations of systematic thinkers in the west from Plato through Hegel and Marx to Carnap and Goodman may be, none of them in fact takes up the challenge of the restrictions upon intellectual assertion leveled by the goddess. Plato (who is most often taken to be Parmenides' heir) after explicitly merging the concept of Being and Non-being to form the hybrid Becoming, throws up his hands in despair of ever understanding what the Master really meant. The cosmological sphere of Being that Plato presents in the *Timaeus*, and which is generally thought to stem from Parmenides, is in fact a version of the sphere that Parmenides describes in the second part of his poem—the part presumably dedicated to a presentation of the way of "opinion," namely

Introduction

—a "likely construction", an "eokos logos." But whereas for Plato such constructions are things one proposes when one has a hint of the truth but cannot prove it, precisely such hints and propositions are, by Parmenides, interdicted.

Parmenides, then, far from being the progenitor of the totalizations and hierarchies of subsequent western thought, and thus the figure most to be assailed at a time in history when western thought is tottering, stands outside of the whole tradition as a monstrance and a castigation against it, out of its origin. Parmenides was a hope that has not yet been fulfilled because his actual proposition has never been seriously entertained or analyzed with any particular depth.

The connection between the Parmenides poem and the cultures of India has been remarked upon. Something like Parmenides' sphere of Being exists as a symbol for the Buddhist's principle of the Dharmakaya; and, the expression of the absolute object of contemplation through the denial of a series of predicates is a common feature of Indian absolutism. Ingenious analyses have been offered likening Parmenides' chariot journey—with its axle that gives off musical tones and radiates heat and light—to the processes of Kundalini Yoga. That Parmenides was educated as a Pythagorean is part of the Parmenides tradition. That Pythagoras' doctrine, with its vegetarianism, metempsychosis and general soteriological scheme owes *its* similarity to Indian metaphysics to Pythagoras' personal journeys to the east is part of the Pythagorean tradition. Thus the possibility exists that the Parmenides poem reflects a hidden dimension of cross-cultural fertilization between India and the Ancient Mediterranean.

Another aspect of the Parmenides poem that looks suspiciously like typically Indian approaches to epistemology is its bipartite structure: the poem contains a doctrine of "two-truths," one pointing towards an absolute that cannot be expressed properly in speech but can be acceded to through direct realization, the other a relative, conventional truth—a matter of historical convenience, common sense, or heuristic practice. The doctrine of two-truths is prominent in both Buddhist and Hindu metaphysics. Whether or not Parmenides' poem owes anything to the east in this regard, the division of the poem into two contradictory doctrines deserves comment. A text that argues elaborately for the connection with India and which we discussed in Rhinebeck is Oscar Marcel Hinze's *Tantra Vidya* (Motilal Banarsidass, Delhi, 1979). I will reserve my own remarks on this topic for the supplement to this issue of *Io* where there will be further material dealing with Parmenidean questions.

My translation of the Parmenides poem provides a few new readings but I was mostly concerned to take the text seriously rhetorically as a poem intended to celebrate the goddess Dike. Once again, I reserve comment on specific interpretative aspects of the translation for a supplement to this issue of *Io*.

Translation of course *implies* interpretation, and the question of interpretation/translation was an important aspect of the entire Rhinebeck enterprise. Quasha's poems to some extent "translate" the states of cognitivity internal to their own composition, as well as interact in concrete and subversive ways with the "interpretative" machinery and habits of their readers. Hennix's text "translates/interprets" texts from the logico-mathematical tradition—Frege, Wittgenstein, Brouwer—in part by inserting them in a series which includes Parmenides on the one hand and the Japanese *Hekigan Roku* on the other.

L.E.J. Brouwer

After Parmenides, a second personage studied by us in Rhinebeck was the modern Dutch mathematician, L.E.J. Brouwer. Brouwer is the founder of the Intuitionist School of Foundations of Mathematics. He was himself a contemplative and a mystic, and the inspiration and motivation for his mathematical work remained throughout his career inseparable from his spiritual concerns. Brouwer is a major influence on modern mathematical theory, both behind the scenes and center stage, though his work is not frequently studied today in academic circles even where intuitionism of a certain type is in favor. He is, however, a major influence on both A.S. Yessenin-Volpin and Christer Hennix. The appendix to this introduction includes several of the key passages from Brouwer's work that express his views on the contemplative life. I will attempt, in due course, a conjecture as to the connection between mathematics and contemplation his work suggests.

In order to see Brouwer's position in mathematical thought, it is necessary to grasp how the intuitionist school compares in outlook with other mathematical philosophies. I will sketch these views very briefly, and refer the reader to the introduction and opening selections in the anthology of essays edited by Paul Benacerraf and Hilary Putnam, *Philosophy of Mathematics, Selected readings*, (2d edition, Cambridge University Press, 1983) for further discussion.

It is often remarked that there have been three distinct approaches to the question of the ontological status of mathematical objects in

Introduction

modern times:
1. classicism (or platonism)
2. formalism (or conventionalism)
3. intuitionism (or constructivism)

with constructivism playing a part in formalism.

For the classicist or platonist, mathematical objects are objectively-existing entities. The mathematician discovers these entities and studies their properties in a manner analogous to the way a natural scientist discovers and studies natural phenomena. Thus, the intellectual activities involved in developing mathematical systems are extrinsic to the systems themselves. Natural numbers, ideal geometrical objects, axiomatic structures—in short, mathematical entities in general—are considered to be existences independent of the thinking mathematician. Mathematical language expresses but does not determine mathematical truths, and the entire mathematical universe is conceived to be a hierarchy with logic and set theory at the apex.

In contrast, for the formalist, mathematical objects are purely conventional. They consist solely of written marks and the rules devised by persons for the manipulation of those marks. Mathematical systems have no ontological significance apart from the ontology implied in their being concrete inscriptions of signs. Whereas for the classicist a proof is a demonstration that certain objects exist in an ideal mathematical cosmos, for the formalist a proof is simply an algorithm: a strict procedure for deriving strings of symbols from other strings according to perfectly explicit rules. Any relationship to "reality" that such systems may possess derives from interpretations or "models" of those systems: ways of associating certain strings of symbols with physical or other realities. The intellectual activities involved in developing mathematical systems are perfectly extrinsic to the systems themselves, except in the sense that the activity of devising the rules of procedure is an intellectual activity. The rules once devised, however, can be executed perfectly well by a computing device without reference to human intervention. Mathematical language (the marks and the rules for use) constitutes the whole of mathematics.

For the intuitionist, mathematical objects are constructions of the human intellect: the "creating" or "creative" subject, to use a phrase coined by Brouwer. A mathematical object is intrinsic to the intellectual activity involved in thinking it. It is neither true by convention, nor is it a feature of the objective universe. In fact, in its most radical formulations, both the objective universe and the world of human convention are themselves derivative of the mathematical activity of thinking subjects. A mathematical object becomes real by its being thought, by its being constructed concretely in the mind of a thinking mathematician.

While classicism is in a sense derived from platonic philosophy with its emphasis on the reality of abstract ideas, its modern developments are due more to the mathematicians of the Renaissance and finally to Descartes than to the notions of the Greeks. (For an excellent account of the development of Renaissance mathematics and how it differs from that of the properly Classical period, see Jacob Klein's *Greek Mathematics and The Rise of Modern Algebra* and Don Byrd's essay in this issue.)

Modern formalism is associated mainly with David Hilbert. Intuitionism derives ultimately from Kant's apriorism, but gets its major impetus from Brouwer. Brouwer's approach to mathematics was based as we have noted, upon a perspective that was decidedly "mystical" or "contemplative" in origin. Brouwer's Ph.D. thesis, *Over de Grondslagen der Wiskunde*, in which many of his major mathematical notions receive their intitial formulations, was composed simultaneously with his *Life, Art, and Mysticism*. The two texts were intended to complement each other, though his thesis advisor forced him to remove the passages from his thesis which would have provided the links between them. Brouwer never retracted his mystical views, however embarrassing they proved to his tutors and subsequently to his disciples.

Brouwer's intutionism begins with a severe critique of classical mathematics on numerous points — points which were later taken up by Hilbert and his followers and incorporated in the conventionalist program. Later too, Brouwer's intuitionistic reconstruction of mathematics was rewritten by his disciples without grounding the system upon the fundamental intuitionistic acts that had been their origin. In other words, much of what is presently called "intuitionism," though derived indeed from Brouwer's alternative methods for constructing mathematics, is intuitionism in name alone in that it does not seriously depend upon the concrete acts of intuition in the mathematical thinker's mind.

Brouwer viewed mathematics as being built up from certain profoundly private internal acts of the thinking psyche. These acts take place in the loneliness of the thinker's internal life and constitute the inner gestures performed between the self and its deity. Mathematical

Introduction

intuition is thus fundamentally independent of language —independent even of the disambiguated language of mathematical thought —and profoundly removed from the language of everyday. Mathematical language functions to help the thinker remember his thoughts or to aid in the stimulation of such thoughts in others, but is finally inessential to the nature of mathematical truth itself.

The controversy between the three main schools of mathematical thought of course touches many philosophical topics, but let me confine myself to the following two:

1. the question of the ontological validity of natural science;
2. the link between intuitionistic thought and contemplative modes of enlightenment.

Natural science aspires to validation through a certain hierarchical organization. The social sciences and "sciences of man" must depend upon the biological facts pertaining to the human species as discovered by biologists. Biology in turn grounds its analyses of biological phenomena upon the chemical substrate of those phenomena; and chemistry has achieved "certainty" through its dependence upon quantum physics. Physics is the ideal and basis for all natural science in that its findings are expressed in precise mathematical language. Indeed, the truth claims of the physicist are really limited to the formulae and equations which depict the regularities discovered in the experimental data. Thus, if physics possesses determinate ontological content, that content both determines the ontology of the other sciences and rests upon the ontological import (or lack thereof) of the mathematical language in which it is expressed.

Of course this view of things is over-simplified and not without its challengers, and would need much qualification to be made precise —there are numerous biological and sociological concepts which are not derived from the position in the above hierarchy: concepts of structure, function, pattern etc. But all such concepts apply to physical systems which in turn have their analyses on the next hierarchical level. If the independent concepts are to be given ontological significance without recourse to this founding level, the entire edifice of physical science loses its coherence, and ontological principles outside of physical science must be sought to ground these structures. Again, certain phenomenologists argue that the hierarchy sketched above itself grows as "regional ontologies" or as derivative structures from more primitive domains. But in such cases the ontological significance of science is sought outside of the domain of science. The point here is simply that if the edifice of science as a whole demands recognition as positive knowledge, then it must be able to provide its own ontology. In any case the ontological significance attached to the mathematical language in which the base of the system is expressed will determine just what the ontological significance of the sciences are. It will be argued by some that, presently, science proceeds not as an organized hierarchy of disciplines, but as a multiplicity of heterogeneous activities with various relations to each other and to the world. But this condition of incoherent multiplicity follows from the abandonment of ontological finality as a desideratum, a consequence of precisely the formalist philosophy. At the very least what can be said is that the current state of things rests upon a philosophical outlook which itself is far from unanimously defended.

In any case, the ontology (or lack thereof) of natural science depends upon the nature of the "truth" inherent in mathematical language, or the nature of mathematical validity itself. If there is NO CONSENSUS on the nature of mathematical language, any consensus the physicists arrive at regarding their own theories must be considered indecisive. The many assertions that the ontological meaning of the physicists' theories can well remain indeterminate (since it is the pragmatic, i.e. technological, efficacy of these theories that really backs up their truth claims) are all well and good IF one is willing to admit the reduction of all science, including theiretical physics, to the province of technology. This admission would entail that there is no "pure" science at all, and that to look to science for anything more than technical means for attaining practical ends, must be renounced.

We turn now to a discussion of the link between intutitionistic philosophy and contemplative concerns.

We will see the intuitionist in two quite distinct though related attitudes. One of these is in fact not incompatible with at least certain formalist-constructivist possibilities and has a reasonable number of adherents (including, in a sense, Yessenin-Volpin, though Yessenin-Volpin goes much further than other formal-intuitionists in emphasizing the concrete realization of construction methods. Also, Yessenin-Volpin's work is ontologically *anarchic* rather than, as is the case with formalism, *indifferent*.) The second attitude, though clearly the attitude of Brouwer himself, has few vocal adherents in the mathematical community. Let us call the first attitude *formal intuitionism* and the second, *intuitionism proper*. Formal intuitionism and intuitionism proper do not differ necessarily in their propositional content, but

Introduction

rather in their ontological stance. Thus, I will sketch first formal intuitionism, and what I say here can be assumed to hold for intuition proper as well, though not conversely.

Formal intuitionism coincides with classical mathematics when dealing with finite mathematical objects. But the main developments of modern mathematics have explored the rich domain of infinite or "transfinite" magnitudes and orderings, so the coincidence is in fact relatively trivial. Formal intuitionism denies the law of the excluded middle for infinitary systems. And this denial does not refer in particular to the demand that the value "undecidable" be admitted as an alternative. In classical mathematics, a proposition is *undecidable* or *unsolvable* if the application of the procedures of the system yields NEITHER the values "true" nor "false", when carried out. Such propositions are said to be *independent* of the formal system chosen and they may therefore serve as *axioms*. Hence, the Russell Paradox in naive set theory can be considered undecidable, and in certain interpretations of the "Liar Paradox" the liar-statement is judged to be undecidable. But for the intuitionist, since an object cannot be said to exist until it has been constructed, nor can it be denied that it might exist until it has been shown that it cannot be constructed — there will be objects whose existence can neither be affirmed nor denied. These objects will also not necessarily lead to paradoxes. The point is not that the existence of such objects cannot be decided formally, but rather that their existence HAS NOT YET been determined. Time distinctions thus become an important part of Intuitionist thinking, especially, in the system of Yessenin-Volpin.

A frequently cited example:

Consider the decimal expansion of Π. A procedure has long been known for indefinitely extending the sequence of digits defining the value of this irrational number. Today, this procedure is being carried out to ever-further decimal places by computers, so that progressively the value of Π will grow more and more lengthy and more and more "precise". Now consider some arbitrary sequence of digits (a,b,c, ... n) that has not as yet turned up as a particular segment of the sequence of digits constituting the decimal expansion of Π. We ask whether the proposition:

P.1.   "(a,b,c, ... n) belongs to the decimal expansion of Π"

is true or false? According to a classicist, this question has a unique answer that already exists: it just hasn't been discovered yet. But the intuitionist points out that, although as soon as (a,b,c, ... n) has been found, the value "true" (or not false) can be assigned to the proposition, since the expansion of Π does not terminate, the procedure could be applied indefinitely without our having a proof that P.1. is either true or false. There never will come a time when we will have generated a sufficiently long enough decimal expansion to decide the question negatively. The intuitionist therefore argues that the answer to our question does not at the present time exist and that the classical position has been refuted. We cannot know anything about the properties of such a sequence until we actually construct it.

The example of the decimal expansion of Π is not trivial. The question it raises infects the entire question of the "existence" of irrational numbers (which are understood as non-repeating non-terminating decimals) and through these to the understanding of the structure of the mathematical continuum, i.e. the points laid out on the "real line". For Brouwer, the continuum is not a unique concept with eternal properties to be established once and for all. Its nature varies with both the specific details of its construction and the precise degree to which procedures of its construction, at a given point in time, have been carried out.

Formal intuitionism thus amounts to a restriction on logical rules when applied to infinite objects. As such, it can be taken over whole cloth and given a formalist interpretation. Such in fact was carried out by students of Brouwer; and, today, when logic itself has been generalized within Category Theory, Intuitionistic Logic and Intuitionistic Set Theory appear as formal variants of mathematical structures, as legitimate within logic as non-Euclidean geometries are within modern geometry.

But what is the relationship between formal intuitionism and intuitionism proper — intuitionism as invented and presumably practiced by Brouwer? And how does this relationship bear on the possibility of a rapprochement between mathematical thought and contemplative consciousness?

Brouwer's intuitionist mathematics cuts down severely on the universes (species) of mathematical objects. The motivation for doing this was to provide a foundation for mathematics. In prosecution of this aim, Intuitionism is in competition with the project initiated by Frege, watered down by Russell and Whitehead, and brought to a head by Hilbert and his school (including the young Gödel), to reduce all of mathematics to a *finitary* structure such as, for example, a finite set of

Introduction

axioms. Frege had actually proceded along intuitionistic or semi-intuitionistic lines, identifying logical norms with general principles which can be *effectively carried out* (though he is commonly thought to belong to the classical camp); Hilbert, working formalistically, also viewed logic with a constructive interpretation where, in particular, the concept of *decidability* was emphasized: you stipulate the rules of logic explicitly so that they can appear as algorithms. But the goal of Hilbert's project was to reduce all of mathematics to a finitary system called *metamathematics* (or *proof-theory*) and, as a by-product, to obtain a *consistency proof* for classical arithmetic by the use of logical principles even more restricted than Brouwer's! In fact, this is precisely Hilbert's "second problem." In contrast to both of these, Brouwer thought all intellectual activities carried out by means of notational systems (whether those systems are derived from ordinary language as in philosophy or inscribed in the rarified notation of the formal logic of Russell, Whitehead et. al.) — that all intellectual disciplines of this sort are in fact initiated by human conventions for human convenience. Mathematics, however, is based on fundamental and inalienable intuitions that precede all logical or linguistic norms. The foundations of mathematics is not to be sought in the logic of formal systems at all. Mathematics is to be founded by linking it to fundamental intuitions of thought-processes of the thinking subject. Moreover, these intuitions of thought-processes are at the basis not only of mathematical thinking, but of consciousness itself. Brouwer's speculation founds mathematics upon specific modes of human intuition, while at the same time finds human consciousness itself to be a form of mathematics (mathematics being understood precisely in the intuitionistic sense).

The fundamental intuition that grounds both human consciousness and mathematics, Brouwer calls the awareness of "a move of time." A move of time is a minimal element of consciousness. It is similar in nature to the notion of a "point-instant," (familiar from occasionalist thought in its various forms—Islamic, Jewish, Buddhist, Cartesian, Whiteheadean, Russellian, etc.) i.e., the notion that time (and the material objects, or moments of awareness within time) consists of a linear continuum of point-like moments of infinitesimal duration, which vanish as they arise and are replaced spontaneously by successor moments. This concept of time, however, abstracts from dynamic temporal experience and reduces the enigmas of temporality to the problem of the linear continuum. For Brouwer, the linear continuum itself is to be built up from the experience of time, so such a reduction is impossible. What is required is that certain fundamental properties of the continuum (for instance, that neither temporal continuity nor discreteness are derived from each other) must already be present in the temporal experience from which they are to be extrapolated. Brouwer accomplishes this by asserting that the fundamental element of temporal consciousness already has an irreducibly complex structure: it is not a point-instant, an indivisible, durationless unity, but a "twoity"—a dyad with the following characteristic: each moment of consciousness spontaneously splits into two parts—1, the trace of itself surviving as a spontaneous memory; 2, the succeeding moment. This new moment repeats the process, similarly splitting into the trace of itself and its own successor. Both the dynamic character of time and the way consciousness is bound up with temporality are thus built into the fundamental concept. The duality, continuity/discreteness, is also established thus at the foundation.

The discovery of the twoity as the fundamental entity is to be an act the thinking subject must perform for itself. It involves a deliberate concentration of consciousness on its own streaming, and a concrete wresting of this fundamental abstract structural core. The twoity and the succession of twoities are to be discerned as the "empty" substratum for all experience by an act of self-intervention that the reader may feel, as I do, is profoundly connected to the act of "turning into oneself" that Brouwer recommends in *Life, Art, and Mysticism* (see Appendix II). Brouwer does not explicitly draw this connection, but whether or not his mystical speculation will be experienced as relevant to his mathematical thinking depends upon our providing that link. Indeed, whether or not his demand for the restriction of the use of the law of the excluded middle can be connected to his founding of mathematics on radical temporal experience may depend entirely upon whether or not one undergoes that radical temporal experience. In any case, from the procession of empty twoities, one proceeds to develop the elementary series of mathematic thought.

Now the procession of twoities when it becomes explicit for consciousness founds the construction of the conscious intellect, yet at the same time, the procession of twoities itself is inherent in the subject. It is through processes directly analogous to the development of explicit mathematical thinking that consciousness builds up a world. The link between mathematical thinking and the development of consciousness (what Hennix will call attention to as the link between *semeiosis* and *cosmosis*) is the notion of a series. Consciousness experiences two species of series: one, it discovers as causally-related temporal sequences; the other, it constructs into things—objects with durative identity. A causal sequence requires that the order relationship among the members of the series be preserved, while an object is coalesced from a series of impressions whose members may be considered independently of their position in that series. Brouwer writes in "Consciousness, Philosophy and Mathematics" (1948; *Collected Works*, p. 1235):

> An iterative complex of sensations, whose elements have an invariable order of succession in time, whilst if one of its elements occurs, all following elements are expected to occur likewise, in the right order of succession, is called a *causal sequence*.
>
> On the other hand there are iterative complexes of sensations whose elements are permutable in point of time. Some of them are completely estranged from the subject. They are called *things*.

Our belief that event X is the cause of a subsequent Y which in turn is the cause of a third event Z, depends upon the temporal order of X, Y and Z. If Z precedes X, it normally cannot be the cause of it. But the sequence of momentary impressions of an object O can be recalled in any order we choose without disturbing our belief that the object O is the same entity, perduring throughout its various momentary appearances. Causality and thinghood are thus both brought under the single concept of a series, and the latter concept is derived from a fundamental intuition of time. The world is built up by the soul as a complex concatenation of both kinds of such series.

At this point I refer the reader to the passages from Brouwer's work included as an appendix to this introduction. These passages bear upon the notion of the sequences of twoities, and the relation between Brouwer's intuitionism and his mysticism. First, Brouwer's statement of "The Two Acts of Intuitionism" (Appendix I). These passages were repeated in subsequent works and stand as the founding statements of intuitionist thought. They will be followed by excerpts from *Life, Art, and Mysticism*. Both sets are presented as selected by Hennix and published as excerpts in his text *Brouwer's Lattice*. *Brouwer's Lattice* was "required reading" at the seminars and the excerpts represent what Hennix wished to present to us of Brouwer's mystical thought.

Charles Stein

Twoity

There have been very few commentators, besides Hennix, on the connection between Brouwer's contemplative and mathematical views. It has seemed difficult to many to see how Brouwer on the one hand derives arithmetical systems from his twoities, and how, on the other hand, the system derived from twoities and series of twoities link to his mystical ideas.

If the serious business of the soul involves a turning into the self and a turning away from "the sad world" of constructed, willful, public reality, as Brouwer maintains in *Art, Mysticism and Life*, how can the creation of a mathematical universe facilitate this "turning into the self"?

Tentatively we can conjecture that Brouwer's construction of mathematics provides him with the possibility of exhausting the activity of his own intellect in a way that liberates him from the cunning and deceit of the "Sad World." Brouwer's intuitionism provides, at every step, links back to "the Creative Subject's" own activity in constructing its cosmos and provides a system of procedures which makes it possible for anyone whose motivations are congruent with his own to retrace the path of humanity's intellectual catastrophe (as outlined in the excerpts given in our appendix), and, at the same time, to avoid both the absolutism of classical mathematics and the nihilism of the formalist school.

I will develop this conjecture by considering how Brouwer escapes from the deceptive consequences of the concatenation of the Sad World. (I use Brouwer's phrase "the Sad World" to refer to the world of human experience and culture as built up from serial concatenations.) This will involve first of all an interpretation of how both arithmetic and consciousness are derived from the series of twoities.

The generation of the twoities in Brouwer's system can be interpreted as developing along a pair of axes: the reiterated series of empty twoities proceed concretely in time along a horizontal axis; processes of intellectually constructing further series of twoities (or multiplicities of series of twoities) occur along the vertical axis.

The horizontal series of twoities, developing in time, constitute a concrete experience of emptiness—a fundamental experience of time to which the Creative Subject returns *ad libidum*, as a way of "emptying out" its consciousness, at any point in the process of construction. The series of empty twoities forms a fixed series, a bottom line to which

Introduction

consciousness is always free to return. It is the basis of all constructed existence, the reiterable fundamental intuition grounding and re-grounding both thought and the world grasped by perception and thought. At every moment it is independent of all concrete content, yet, because it is wedded to a specifically subjective experience of time itself, it is the most concrete of things.

The vertical series of constructions represents successive generations of indices for chains of positive cognitive acts, perceptions, and mathematical constructions. The complexity of our world and of our thought, and our bewilderment before both—all derive from the interplay of these vertical constructions proceeding in forgetfulness of their purely derivative character.

The vertical axis is built up as follows:

Arising from any twoity along the horizontal axis, conceive a series of abstract constructions. At the position immediately above any selected twoity, find a conceptual copy of that twoity together with a copy of the two unities of which the original twoity is composed.

These two unities, it will be remembered, represent a pure moment of sentience (divested of all content) which has spontaneously split into a successor moment and a trace of itself. The abstract copy of the original twoity together with the copy of the two unities constitutes a new twoity: a new bi-unity, an abstract representation of the original pair of unities of which the original twoity was composed, and the conceptual combination of these two unities into a new bi-unity. This process can now be iterated, each new abstract twoity being composed of the two prior unities together with their unification into a new unity, constituting once again a new bi-unity. Each successive iteration of this process grows more complex, in the sense that the content of each consists of larger and larger complexes of twoities and unities. And the series of ever greater complexes of twoities and unities will be able to serve as an index for ordering other sequences of objects, perceptions, or constructions. The series of twoities, ever growing in complexity, serves as a construction of a natural number series or even the continuum of real numbers itself.

While the horizontal series takes place in the concrete real time of the Creative Subject, with each twoity existing only in its own moment and surviving only as its own trace in the successor twoity, the vertical series generates an atemporal conceptual space, in which each successive abstract twoity enjoys a kind of hallucinatory pseudo-existence.

Each successive stage in the development of the "vertical" series must take place within the real-time existence of the creative subject, yet it is precisely the property of this "ascending" series of abstractions to neglect this temporal character, and to generate a symbolic space in which predecessor twoities continue to exist as successor twoities are formulated.

Though we have been referring to the original succession of horizontal twoities as a series, as a matter of fact, the possibility of anything existing in series only emerges with the vertical series. The horizontally-extending emergings of twoties perpetually perishes, leaving only its traces in successor twoities. It only becomes possible to identify the horizontally emerging twoities *as* a series after we have indices for them. And this becomes possible through the fact that the ordering of the vertical series can serve as a system of indices. (The possibility of the construction of the vertical series, however, derives from the fact that the horizontally-extending twoities have a dyadic structure: the individual point-moments of temporal consciousness are in fact bi-unitary, for it is only by abstracting from the possibility of a trace of a predecessor moment that an abstract series of ascending twoities can be formed.)

The vertical series of twoities can, as we mentioned, be treated as a natural number series: a set of indices whereby other series of otherwise absolutely transitory phenomena in the experience of the Creative Subject can be linked to a conceptual space which enjoys quasi-independence from the uncompromised evanescence of the temporal domain. Thus, the two species of series (which Brouwer identifies with objects and causal sequences) are concatenated by consciousness by linking them to members in the vertical series of twoities which, in the form of natural number series, serve as indices for their members.

The existence of both objects and causal sequences are capable of being cognized by the Creative Subject because they can be indexed by the members of the vertical series of twoities.

But because the entire vertical series of twoities has no concrete existence, the Creative Subject is at every point of its existence free to "drop down", as it were, to the horizontal axis and recover the emptiness of its own actual condition. The vertical series of twoities (and the causal and objective series for which it serves as an index) only exists as long as a concrete Creative Subject sustains it and no longer. The entire construction of the Sad World is thus seen as being at every point capable of being collapsed back into the empty but concrete continuum of the creative subject's only authentic life.

Introduction

This possibility of "dropping down" to the horizontal axis at any point in its life allows the Creative Subject to remain free from ensnarement by the construction of intellect and consciousness, free at every moment to resume its silent communion with its spiritual ground.

In Hennix's work, the concept of the Creative Subject, taken over from Brouwer, together with this bi-axial construction, gets explicit development. In his "Theory of The Creative Subject" (page 383) the atomic conceptual acts called "noemas" correspond to Brouwer's series of temporal twoities, while the vertically-rising series of denotational connections correspond to Brouwer's vertical axis of intellectually constructed twoities. But before looking at Hennix's work we need to sketch some of the main ideas of Yessenin-Volpin. We will also look at two further topics of general interest that play key roles in Hennix's thought: Category Theory and The Procedure of Diagonal Substitution.

A.S. Yessenin-Volpin

Alexander Sergevich Yessenin-Volpin is a Russian refugee who has recently become a naturalized American citizen and who presently lives in Boston. He is perhaps best known as a co-founder of the Soviet Human Rights Movement (together with Sakharov), but he is also the founder of "The Ultra-Intuitionist School in the Foundations of Mathematics." An outline of the program for this school is contained in an article published in Kino, Myhill, Velsey: *Intuitionism & Proof Theory* (Amsterdam, 1970). Though it is far too subtle and intricate for me to summarize fully in these pages, I strongly recommend it to the readers of *Io*. It makes few demands upon the reader in terms of formal preparation, develops its own symbolism quite lucidly, and is thus a text which can be read by the interested novice.

To give a sense of who Yessenin-Volpin is as a human-rights activist, I quote here from a biographical, editorial introduction to an article by Yessenin-Volpin from *The Humanist*, January/February 1973. "After many attempts to leave the Soviet Union failed, Aleksandr Yesenin-Volpin, distinguished mathematical logician, was finally granted an exit visa in May of 1972....

"Dr. Volpin had been a key figure in the democratic movement in the Soviet Union, and, as a result, he had been constantly harassed and intimidated by the government. During Stalin's era, he was committed to a mental institution for writing two anti-Stalinist poems.... After Stalin's death he received a pardon. He edited the section on Mathematical Logic of the journal Mathematica.... In 1959, he was invited to participate in an international symposium in Poland but was denied a visa. He was imprisoned in 1959 and put in a special mental hospital in Leningrad, but was released in 1961...

"On December 5, 1965 (Constitution Day) he participated in a demonstration on behalf of Siniavski and Daniel in Pushkin Plaza, Moscow, carrying a sign reading "RESPECT THE CONSTITUTION! He was arrested, but released.... In February, 1968, he applied for a visa to travel abroad. He was, for the fifth time, incarcerated in a mental institution. A letter of protest from 95 mathematicians and scientists helped to secure his release."

Yessenin-Volpin calls his program "an anti-traditional program" and it is based upon a putting into doubt of at least eleven principle assumptions usually left unexamined in classical mathematics. Rigor in mathematics generally develops by making explicit assumptions which hitherto had been either concealed or taken for granted. Making such assumptions explicit may have the destructive but salutary effect of pointing out contradictions hidden in what had seemed to be a secure system. But such destructive discoveries most frequently lead to a deepening of mathematical understanding and the opening of fresh ground.

Ground-breaking criticism of this order, however, frequently meets with a two-fold resistance: first, the desire not to shake unnecessarily the foundations of systems and practices apparently in good use; second, the seeming self-evidence, triviality and harmlessness of the assumptions exposed. Novices in mathematics complain that much discussion in foundations amounts to quibbling over what seem to be pointless differences. Yet it is precisely at the level of seemingly unquestionable assertions that the structure of our unconscious cognitive commitments reveal themselves. For this reason, Yessenin-Volpin operates on the principle that everything should be open to question.

I will not be able to examine all the points that Yessenin-Volpin challenges, but will sketch out the implications of a few of them. First of all, he rejects the uniqueness of the system of natural numbers. The natural numbers are not a system of entities existing in a single form for all time. They must be constructed concretely for each context of use. And different constructions can be brought into comparison. Yessenin-Volpin shows that it is possible, for instance, to disagree about the nature of an infinite series of natural numbers: the classical concept of infinity holds that the natural numbers form a completed totality. Yet it is possible to hold that such a totality does not exist, i.e. that the natural

Introduction

numbers only extend as far as it is effectively possible to compute them. To give an *informal* example: a number such as $[10^{12}]$ to which no one has ever counted, since it is not a *humanly* (non-computer-aided) feasible number, can be defined as "playing the role" of an infinite number in a humanly-feasible substructure of arithmetic. Since it can be shown that it is possible to construct number systems according to these two different concepts of the meaning of the term "infinity," it is thereby informally demonstrated that the natural numbers do not form a unique system, i.e. that axioms of limitation of size in arithmetic are *independent* from each other. Yessenin-Volpin generalizes this idea presented as systems in which two (or more) number systems are employed concurrently, and specific rules for embedding one inside the other are established. (The uniqueness of the natural numbers is a main feature of the classical belief that mathematical entities are objective and independent of our constructions. The recognition of the plurality of number-systems actually is a return to a way of thinking about numbers which was current among the Greeks. See Jacob Klein's *Greek Mathematics and the Rise of Modern Algebra*.)

Yessenin-Volpin also doubts the principle of mathematical induction and the assumption, which he calls the "locality principle," that, "If the axioms of a formal system are true and the rules of inference conserve the truth, then each theorem is true." Induction is the principle which allows the classical mathematician to assume that all the natural numbers "already" exist without having to be constructed or to prove that every natural number is finite. (0 is finite and if n is finite so is n+1. Then all natural numbers are finite.) Induction allows the mathematician to assume that properties that hold for numbers hold for their successors. Volpin shows that the locality principle and induction require each other: to derive the theorems from an axiomatic system, you must assume that induction holds; but, to demonstrate the truth of induction, you make use of an axiomatic system and in particular the locality principle. The two principles form a vicious circle and therefore ought not to be used at the *foundations* of mathematics.

The above considerations do not mean that there cannot be explicit and well-defined employments of axiomatic systems. They only imply that such systems do not exist as absolute constructions, guaranteed to formalize truth without any prior considerations as to the specific aims and conditions of their being set up. In other words, Yessenin-Volpin requires that whatever mathematical system one employs, the aims and rules for its constructions must be specified explicitly beforehand. The terms, demonstrations, and deductions of a classical system are only what he calls "termoids," "demonstroids" and "deductoids": i.e. they are demonstrations and proofs which require further comment showing that their construction has been carried out in comformation with the explicitly stated "proto-theories": the definitions, rules and aims that are established before the axioms and rules of inference are laid down. In particular, in order for a formal proof to be "convincing" all term*oids* which occur in such a proof must be shown to be *terms*, i.e. signs which demonstrably carry denotations.

The fact that not only definitions but AIMS must be stipulated before a mathematical system gets under way, radically relativizes and concretizes the entire practice of mathematics. It effectively destroys the insularity from social, personal, and ethical contexts that mathematical thinking has traditionally enjoyed. For if it is required that aims for mathematical thinking must be made explicit, then very general considerations about the nature of aims and means become directly connected to mathematics.

The Logic of the Moral Sciences is a treatise stating precisely the ethical principles which he recommends. As an ethical theory it is distinguished from other formal ethical theories by the fact that it does not reduce general ethical principles to a formal language assumed to be valid independently of ethics. Rather, it attempts to think through a set of principles which, though far broader than those involved in specific mathematical formalisms, nonetheless are required in order to specify the conditions under which mathematical systems are to be set up. These principles are concerned with the nature of aims and rules, and thus find applications in other areas of human endeavor where aims and rules are relevant. In particular, Yessenin-Volpin develops his ethical theory in relationship to matters of jurisprudence. But it is clear that an extension of these ideas could be developed, say, in aesthetics, or religious practice.

Two aspects of Yessenin-Volpin's system deserve special comment. First, Yessenin-Volpin defines logic itself as *the science of avoiding errors*. Errors are understood relative to aims. Logic then is a discipline that must be developed locally in every case relative to specified goals. It is not an aprioristic science that legislates demands about what we should or should not think without regard to the purposes for which we ourselves in each case employ logic. The consequences of logic are thus only binding relative to our own intentions.

Introduction

Second, relative to situations in which we cannot afford to make errors—where, in other words, logic is indispensible—it is propitious that we should wish to banish from our considerations acts of "faith." Faith here means trust in assertions adopted without "sufficient reason," i.e. without proof. But proof does not mean in general the algorithms by which theorems are generated from axioms according to inference rules. A proof is rather "any fair means of establishing incontestability." "Incontestability" is not defined independently of ourselves as concrete users of logic; and, even in the face of a proof which we are unable to contest, we retain the freedom to neglect it. Though Yessenin-Volpin's system demands a level of rigor unprecedented in any prior mathematical philosophy, the point of the rigor is not to establish an absolutely binding universe of truths, but to secure explicitly our freedom in regard to the employment of our reason.

Another interesting feature of Yessenin-Volpin's ethical system is the concept of a "logic of confidence". Occasions arise where it is not possible to banish faith utterly from our considerations, but where it is appropriate to trust in the assertions of others. The logic of confidence thus involves the syllogism, I have confidence in X; X asserts Y; therefore, I am confident in asserting Y. Acts of confidence are capable of being made explicit, and rules for accepting and denying confidence can be stated.

The relativization of logic also involves calling attention to what Yessenin-Volpin calls "collations" and "tactics of attention and neglect." Collations are connections between things. They are of two kinds: identifications and distinctions. These are taken as the atomic actions of our reason. They are things we do. Thus "identity" and "difference" are not fundamental ontological categories: they are consequences of our activities undertaken in concrete contexts. Similarly, the rules by which we follow a given rational procedure depend upon "tactics" by which we follow the rules and other tactics by which we give ourselves the permission to neglect certain aspects of the situations in which we are operating. Both these species of activities must be made explicit as a pre-requisite for *any* reasoning about them.

Finally, the relativization of logic and the insistence that all steps in a mathematical activity be explicitly justifiable (in the last instance by the law of sufficient reason) imply that the temporal ordering of our logical activities is non-trivial to their results. It matters whether or not a given assertion has as yet been demonstrated. Thus a "tense-logic" must be developed making all these relations explicit.

All of these matters are sketched out in the paper included in these pages.

Category Theory

Category theory is among the most abstract systems of mathematical thought that exist to date. The "categories" of category theory are themselves whole mathematical systems; thus, the sets in Zermelo-Frankel set-theory are objects in a category, topological spaces form a category, linear algebra is a category, group theory is a category (category theory generalizes the well-known concept of a semi-group with unity). In other words, category theory generalizes over and draws its concepts by abstracting from lower-order mathematical disciplines, much as traditional algebra generalizes over arithmetic or topology generalizes over geometry.

Category theory, however, reaches such a level of abstraction that even logical constants and most of the logical rules applicable in lower branches of mathematics are themselves susceptible to analyses and thus to variations.

Though category theory generalizes over lower disciplines, it remains in its concepts and terms quite free of the disciplines it "categorizes." It may more fruitfully be thought of as an independent body of concepts from which other disciplines can be *derived*, than as an abstractive generalization from those disciplines.

Initially, the inventors of category theory hoped to provide a foundation for mathematics alternative to that provided by set theory, but that hope was disappointed because of the theory's extremely high level of generalization. Eventually, Lawvere and Tierney elaborated the notion of a *topo*s: a topos adds additional structure to the notion of a category and makes it appear to be suitable for the founding of mathematics. The topos concept, however, turned out to involve intuitionistic principles, (formally intuitionistic principles, that is), so one had the rather paradoxical situation of being in possession of an intuitionistic theory of the foundations of classical mathematics. It is this circumstance upon which Hennix seizes in his work.

The entities with which category theory concerns itself are called "objects" and "morphisms." An object may be anything at all *conceived of as an object*. Thus, the objects of a category might be sets or natural numbers or topological spaces; but, they might also be states of consciousness or ordinary items of the every day world. The "morphisms", which are generally indicated by arrows connecting the symbols for

Introduction

objects, stand for operations which are performed on, among, or between the objects. If the category is set-theory, for example, the objects would be the sets and the morphisms, the set-functions such as membership, intersection, union, and the like.

In addition to possessing objects and morphisms, in order to qualify as a category, a system must enforce certain rules. Morphisms must be capable of "composition": that is to say, a transitive relation between morphisms must obtain, such that the morphisms can be performed in sequence: if an arrow f connects an object A to an object B

A————f————▶B, and another arrow g connects object B to object

B————g————▶C, there will always be a third arrow h called the *composition* of f with g connecting A to C. The diagram

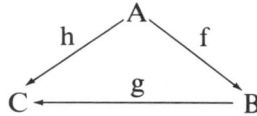

is called a "commutative" diagram and is used in category theory to render, in a compact and intuitively obvious notation, the relations between the objects and the morphisms. The composition of arrows must be transitive except in the case of the "empty category" : the category which has no objects and no arrows. If a mathematical system does not exhibit transitivity, then it cannot be generated within category theory, i.e. an intransitive system is represented by the empty category.

In standard developments of category theory the empty category has little application. As most mathematical systems affirm transitivity, the empty category is developed within category theory mostly to complete the symbolism. Yet in Hennix's use of category theory, it is precisely the empty or "almost empty" category that proves to be the richest and most rewarding of study, since his concern is frequently with *intransitive* structures—structures which thus occupy what he calls "the interior" of the empty category (interpreted as a set).

Since category theory deals with principles held in common by different systems, special processes are required to establish connections, transformations, and relations between different categories. Such processes are called "functors" and, like morphisms, are indicated by arrows, pointing now from category to category rather from object to object.

Among the morphisms that get particular attention in Hennix's work are the "pushouts" and "pullbacks." Pushouts may be viewed as arrows that indicate the generation of theorems from axioms within a category, while pullbacks may be viewed as indicating the reduction of the theorems to the axioms from which they have been generated. (These ideas come from algebraic geometry, in particular, sheaf-theory, and are expressed structurally as "Grothendiech topologies".) A functor that gets special attention is the universal "forgetful" functor which eliminates levels of specification between categories. By a judicious use of pullbacks, forgetful functors and the empty category, Hennix is able to indicate structures which obtain for aesthetic and ascetic processes, usually thought to be incapable of logical presentation.

A topos, as we mentioned, is a category with certain important additional structural features. It was through the elaboration of the topos concept that, as stated above, Lawvere and Tierney were able to initiate the employment of category theory as a theory of the foundations of mathematics. Rather surprisingly, even the Environment E discussed extensively by Hennix below is a topos, or rather contains an unlimited "universe" of toposes.

Toposes contain methods for classifying and indexing the objects of the category, so that it becomes possible in a topos to symbolize a collection of objects, relations and operations together with a system of indices distinguishing and classifying those objects, relationships and operations. It is this feature of the topos that makes it an appropriate structure for the study of the logics implicit in cosmologies, psychologies, and all disciplines where objects must be considered together with their modes of representation.

Diagonal Arguments

The Yellow Book deals throughout with constructions that are known in mathematical language as "diagonal" constructions. These constructions extend formal systems where the application of the operations of such systems to certain values leads to contradiction. A diagonal construction often becomes necessary, for instance, where systems encounter self-reference: if a function is taken as an argument for itself, the value of the function is either un-defined (and the system is incomplete), or the system becomes notationally inconsistent (and the system is over-complete), or else the system must be expanded in some

Introduction

way to accommodate the self-referential case (making use, for instance, of fixed points in metamathematical extensions). The expansion of the system is obtained by first adjoining to it the operation of "diagonalization" and then picking any diagonal formula as a "new axiom". A situation where diagonalization is exemplified usually occurs whenever a formal system is construed as taking arguments for ALL its values (unrestricted quantification). If a free-variable formula is construed with its *free-variable* formulae as ranging over all its values, then there will occur a case where the system will have to refer to itself. In general, where unrestricted self-reference is allowed, paradoxes arise.

Diagonal constructions occur in mathematics where the limits of a given notation are reached, yet where one insists on the possibility of continuing to apply the operations already defined in particular, iterated compositions of operations with themselves. Thus, the creation of fractions to indicate the ratios between whole numbers, the creation of irrationals to indicate roots of numbers not perfect squares, and the recursive development of multiplication from addition or exponentiation from multiplication are all instances of diagonalization.

In modern mathematics, diagonalization plays an important role in the concept of the "transfinite" numbers of Cantor, and the battle over the legitimacy of the transfinite numbers rages over the interpretation of the diagonal procedure that produces them. Again, it is the "diagonal" case in Frege's foundations of arithmetic, the existence of sets which can take themselves as members, that generates the paradox discovered by Russell which constitutes the central crisis in modern mathematical thought; and a very elaborate diagonal procedure is at the heart of Gödel's famous incompleteness theorems. Finally, the entire issue of what is "unsayable" in Wittgenstein's *Tractatus Logico-Philosophicus* hinges upon the diagonal construction which would have to be provided were the process of "picturing" allowed to picture itself.

There is a singular interest attachable to diagonal and self-referential constructions which has been frequently commented upon in both popular and serious mathematical literature, for self-referential situations do not only show up under formal conditions: they may appear wherever a symbolic system is in use, whether such a system is an ordinary language, a philosophical or technical discipline, a biological system of cognition or perception, a work of art, or a system of spiritual exercises. That is, in any system of representation where the free-variable terms of the system have been allowed to range over the totality of their domain, there will occur a diagonal case, a term which refers to the system itself. In each such system, the diagonal or self-referential case will have the appearance of singularity, a moment of confusion, an occasion for doubt, or else will seem a justification for the expansion of the system, for a mystical flight into transcendent realms—depending on the context of the system and the perspicacity of its devisor.

A few examples: the diagonal case in a psychological theory occurs when that theory attempts to account for how the psychology of the theorist affects the psychological theory being propounded (one thinks of C.G. Jung's famous remark, "we do nothing but dream the myth onward" or the problem of transference and counter-transference in psychoanalysis); the diagonal case occurs in anthropology, where anthropologists inquire into the cultural conditions of their own research (and of course, the Lévi-Strauss of *Triste-Tropiques* and *La Pensée Sauvage* comes to mind); the diagonal case occurs in literature in a great variety of forms and wherever authors refer to their own texts (numerous instances of this occur in the poems of George Quasha in this issue); in physics, where experimentalists must give a physical account of their experimental apparatus; and in contemplative exercises, where an instruction to attend to phenomena in a certain way must be applied to the phenomena of obeying that very instruction.

It is generally not observed that the consequences and relevance of each of these cases depend totally upon the concrete circumstances under which they arise. Yet, most frequently, the diagonal case forces a re-examination of the terms of the particular system, often causing doubt, dismay, or grandiose mystification as the case may be. Wittgenstein believed that many paradoxes and puzzles for thought, being bewitchments foisted upon us by this general exigency of our notational systems, could be banished simply by noticing and disallowing diagonal situations from our languages. This observation lead him to dismiss the significance of both the Russell Paradox, which could be handled by simply eliminating self-referential sets from set theory, and the Gödel incompleteness theorems, which appeared to him to involve "a puffed up" notion of proof. (p.132., *Remarks on the Foundations of Mathematics*, 2nd ed.)

It is well-known that historically there have been three typical responses to diagonal situations: rejection of the diagonal case as meaningless; rejection of the system as incoherent; or expansion of the system into a new dimension. In the first instance, one rejects the

Introduction

diagonal case as meaningless according to definition: psychology disqualifies itself from studying the psychology of its own theorizing, a judge disqualifies himself from sitting in trials wherein he may be a party, transfinite numbers are simply denied existence within the mathematical cosmos. But the problem with such a blanket solution is that, though in some cases self-elimination may be appropriate (the case of the judge for instance), in others, the refusal to confront the diagonal case simply begs the question. It may be quite important to know how the psychology of a psychologist affects his psychological theory, or how the physics of a physicist's apparatus influences his experiments. Even transfinite numbers, given appropriate constructions, have their uses.

But if the diagonal case is naively accepted, contradictions in the system arise and the system loses its epistemological coherence altogether: some solution to the Russell Paradox must be accomplished, or else the resulting inconsistencies in set theory render the latter useless for founding arithmetic; if the psychological condition of the psychological theorist determines the content of his theory, the theory is deprived of convincingness; and, if the meditator applies the instruction "refrain from intentional activities" to the activity of meditating itself, he will abandon his session.

Where diagonalization is not abandoned, often a new dimension of the discipline in question comes into view. Diagonalization thus seems to be a moment of creative fertility, wherein the limitations of a notational system are reached and new vistas of conceptualization may be opened. But here the dangers are that of grandiosity and mystification, where irrelevant "acts of faith" (see Yessenin-Volpin below) may be involved. If diagonalization is permitted without restriction, as, for instance, in classical mathematics, we may find ourselves committed to an over-populated cosmos of abstract entities, actually created by the excesses of the notational system.

The later Wittgenstein's relentless assault on Platonic idealism has its roots in his objection to diagonalization procedures, for the belief in the existence of platonic forms has its source in the diagonal situation that arises whenever one inquires about the *existence* not of the entities that fall under a given concept, but of the *concept itself*.

Particularly poignant and anxiety-provoking cases of diagonal substitution occur for the Buddhist meditator, and I will try to show how the example of the Zen student confronting his Koan is analogous to the style of treating diagonal arguments recommended by the ultra-intuitionist program.

Numerous exercises require Zen practitioners to refrain from conceptual thought, to refrain from activity, or to perceive all phenomena and all thought as "empty" or void. But the thought by which the meditation itself is set in motion is itself a concept, and meditation is itself a kind of activity. Meditators must thus wrestle with the diagonal case propounded by the instruction they are attempting to follow, until a break-through, wherein the experience allowing them to differentiate the appropriate levels of application of the instruction, occurs. Diagonalization, in this way, is at the heart of many Zen "Koans" or "Model Subjects" (as they are termed in the *Hekigan Roku*, quoted by Hennix below.) The meditator is forced to address a vivid contradiction which has the structure of a diagonal argument.

Now, Zen meditators notoriously face the irascibility of Zen masters, and Zen masters may refuse the student's solution to the koan for two contrasting reasons: either the student fails to experience the bite of the koan—refuses, that is, to take the contradiction seriously, never actually confronting the diagonal case (such a student is like the formalist theorist who saves the system by rejecting the diagonal case as simply undefined within it—the specific contradiction is resolved, but that is all that has been gained); or else, the student allows himself to grow dizzy with the koan: he experiences a kind of psychic vertigo or ecstasy, allowing the diagonal dilemma to generate hallucinations and grandiose (even transfinite) constructions. His answers to the koan are up in the clouds; his state of mind is inflated and out of touch with the task of confronting the nature of his own mind—the task that is at the heart of the meditative process. The Zen master cuts him down with some direct reminder of his immediate and earthbound condition. Such a meditator resembles the classical theorist, carried away by the platonistic possibilities of his diagonal constructions.

But successful Zen students are able to see the diagonal construction for what it is and pass beyond it. They attain a vantage point that is not confined or defined by the language in which the meditational instruction is couched. They discover in their own experience the very mind that is struggling with the koan, and they do so not by blanking out all cognitive activity or refusing to confront the contradiction, but by exercising their faculties to the limit. They cover the entire space of their cognitive language and step beyond it—not into an ever more elaborate and mystifying hyper-reality of thought construction, but onto the immediate ground of their own activity. They are thereby able

Introduction

to grasp the condition and the structure of the exercise which brought them to their insight.

The acceptable solution to the koan is eminently practical, but by no means reductive: the concrete, down-to-earth mind of the meditator is by no means the mundane confusion of the unenlightened: it is ordinary mind brought into contact with its experiential basis, capable of saturating, at a single stroke, the space of its own activity.

The analogy with the ultra-intuitionist position is this: in dealing with the problem of diagonalization, the ultra-intuitionist steers a middle way between reductive formalism and classical ideational realism. Approaches to diagonalization, like approaches to the construction of number systems in general, are multiple, and tailored to meet specific applications. Because the steps in the procedures of construction must be monitored and ordered, and proofs for the actual completion of the steps must accompany their being carried out, the exact range of application of the propositions of the system can be monitored, specified, and varied according to aim and need. It is unnecessary to banish diagonalization altogether, and it is unnecessary to grant unlimited license to diagonal constructions. Rather, the actual procedure of construction is always kept firmly in view, without arbitrarily and obtusely restricting any constructions which might prove to be of use.

The Yellow Book

Christer Hennix's *The Yellow Book* is a work in progress (Part I of which appeared in *WCH WAY # 5* edited by Don Byrd and Jed Rasula), and is composed in such a way that between any two pages a new page may be inserted (dense order pagination). It is my ambition to provide a commentary for this text (and its dense ordering) at some point in the future, but here I will confine myself to a general description of the work and a somewhat random discussion of some of its salient features.

The Yellow Book consists of excerpts from material written over the last fifteen years, some of which appeared previously as part of Hennix's *Brouwer's Lattice, Notes on Toposes and Adjoints* (1976), and *17 Points on Intensional Logics* (1979). *Notes on Toposes and Adjoints* itself was composed from numerous notebooks to accompany an installation of the (E-) environment, TOPOSES AND ADJOINTS, at Moderna Museet, Stockholm, 4 IX-17 X, 1976. The text was published as an adjunction to *Brouwer's Lattice* and was at once a catalogue for the installation, a specification of its intellectual parameters, and an element in the installation itself. The installation was a work of "Concept Art" more or less as understood and defined by Henry Flynt (cf. his texts included below).

Hennix is presently working towards a re-installation of his environment, which an extended version of the present text will accompany. A description of the non-textual aspects of the installation will enhance our understanding of the present text.

The environment TOPOSES AND ADJOINTS (refered to as $\acute{E}^x$ in the text) consisted of a body of works (1-3) in different media elaborating intertransmutable and even parallel aesthetic and epistemological processes.

1. A TEXT, exhibited on a wall outside the exhibition room of the environment and sold as the catalogue.

2. The installation itself, consisting of a museum room containing acrylic diagrams, monochromatic computer graphics displays, a pair of ultra-black paintings; the diagrams are large-scale wall paintings of presentations of figures implicitly given by the Text, the ultra-black paintings painted by the author and executed with pine-soot as pigment for creating an ultra-black color and applied on highly polished (mirror-like) aluminum square-surfaces, the monochromatic computer graphics ultimately reducible to finite texts defining image-specific computer programs.

3. A continuous, composite sound wave form electronically composed in 1970 and left ongoing from that time to the present.

To participate in TOPOSES AND ADJOINTS actually required spending many hours, not to say days, in attendance at the installation. In attending this programmed environment, one might devote oneself to contemplating the composite sound wave form and its formal relationship with the many ways of reading the graphic material as well as the spatial presentation itself while epistemologically guided by a study of the TEXT.

As mentioned above, the various media of the work were conceived in parallel and are systematically reflected in each other. Each medium—painting, wave form, space, or mathematical text—proposes a certain process of composition and follows that process to one or more limit conditions, for example: a surface of a light-metal mirror with particles of pine soot as the ultra-black paintings are executed until they become totally saturated. Though *discrete* (infinitesimal) particles of soot are applied, this application is continued until an

Introduction

impression of continuous density is reached. At this point, in a sense, an entire world of possible images has been traversed, and yet the final effect—that of a continuously covered surface—repeats and returns to the intitial condition, a blankness or mirror stage in which there is no differentiation of parts (so-called chaotic topological spaces). Sustained contemplation of the black paintings in conjunction with the TEXT, however, recovers all stages of the process: the individual particles of soot become visible again, when perceived as "dissolved" into the printing ink of the text of the formal theory chosen to capture all theorems about a continuum of densely ordered ultra-black soot-particles and for which the paintings act as "models". That is, the "visibility" of the individual soot-particles becomes a semiotical process accompanying acts of self-illumination.

The process of contemplating the composite sound-wave form takes one on a parallel journey through another continuum. The sound wave is composed of three continuous frequencies (sine-waves), selected and structurally organized by Hennix by a technique of eliminating unwanted frequencies from the range of combination tones generated by a selected fundamental tone and the three concurrent sine-waves. The composite sound-wave form is thus generated recursively and continuously; each infinitesimal moment of the composite sound-wave is generated as a recursion performed upon the moment just gone by. Compositional processes have been reduced to a minimum, yet a most stringent format of just intonation is performed uninterruptedly, for which a precise grammar is established. What one hears upon entering the environment E sounds, at first, like a "single note composition". But upon allowing one's attention to be absorbed by the ongoing process of the perception of the single note, the various excluded combination tones and their relationships begin to emerge. If one develops what the TEXT calls a "sustained feeling of awareness" in relation to the sine-wave triad, an infinitely rich world of sound figures begins to emerge in one's consciousness. Further, if one has mastered the conceptual machinery of the TOPOSES AND ADJOINT TEXT, or at least is familiar with the rudiments of Harmonic Analysis (Fourier series etc.), one will find in the contemplation of this composite sound wave form a single SEMIOTIC OBJECT, a philosophical or poetic SIGN whose meanings are generated in the listener's consciousness through the interaction of his/her attention, the structural properties of the composite sound-wave form itself, and the concepts grasped from the study of the TEXT. The indefinite construction of this process is referred to by Hennix as a relation of "homosemeiosis." In particular, the general concept of a "form of thought," given as a space defined by the variables determining its geometry, can be expressed by the same notation as that used to specify the three sine waves of which the sound is composed. As the sound is received as an interpretation of the sound wave in the listener's experience, so a form of thought requires models or interpretations in the Creative Subject's mind in order to be understood. The interpretations for both sound form and form of thought are possible worlds which can meet or even coalesce as a result of a *homotopy operation* (mutual deformations of abstract conceptual spaces).

Now for the Text itself. *The Yellow Book* sketches out domains of abstract thought from various regions of advanced mathematics, in particular topology, topos theory (from category theory), an ultra-intuitionistic version of transfinite arithmetic, and Montague Grammar, interpreted as a generalization of the language of *Tractatus Logico-Philosophicus*. Though the language of these disciplines are employed with formal rigor, the bases and the outcomes of Hennix's speculations by no means leave these theories in place. Just as the soot paintings begin with a blank mirror space and end with a return to the original condition of blankness (by a route that takes the viewer through a domain infinitely rich in specific content and detail), so the theoretical structures of the TEXT build up structures from fundamentally "empty" concepts and collapse all the structures generated back into a recapitulated emptiness at the end, (while at the same time allowing a phantasmagorically rich array of mathematical structures and possibilities to shimmer before the contemplating intellect in passing).

Though the languages of mathematics form the medium of the text's expression, the positivity of mathematical thought is by no means the goal of the enterprise. Quite the contrary: TOPOSES AND ADJOINTS rather diminishes confidence even in intuitionistic mathematical reasoning while at the same time exhausting the mathematical intellect and satisfying its demands. And it does this for ends which are at once mathematical, philosophical, aesthetic, spiritual, political and ethical.

We can see how these ends are accomplished through a brief look at a feature of the TEXT designated as the Theory of The Creative Subject together with its models. Both the notion of the Creative

Introduction

Subject and the structure of its models stem from Brouwer. The Creative Subject is a difficult and somewhat ambiguous conception. It is perhaps best understood as an ideal figure for the individual insofar as he/she engages in mathematical thought. Insofar as the whole of one's experience is involved in the scope of its cognitivity and insofar as cognitivity in general is to be treated according to intuitionistic concepts, the Creative Subject may serve as a first approximation of a model for one's entire cognitive life.

Hennix develops several Diagrams for the Creative Subject and uses them to generalize Brouwer's concept of the fundamental twoities which found all cognitive experience. In one of those diagrams (page 383) Hennix does so along lines similar to those I developed above. This diagram consists of a pair of axes. The horizontal axis represents the series of fundamental intuitions or cognitions called *noemas*. The vertical axis represents chains of cognitions derived from these noemas by means of *arrows* or set theoretical operations. Arrows symbolize *denotational connections* between cognitions. Denotational connections are connections established according to the law of sufficient reason.

There would be numerous applications of the diagrams of the Creative Subject, but the aesthetic principles under which Hennix developed his constructions includes the preference that these applications include a context for working out a foundation for one's personal enlightenment, on the one hand, and a paradigm for the solution of culturally relevant theoretical problems on the other. Both of these demands are fulfillable. In applying the Diagram of the Creative Subject (the diagram on page 383 mentioned above) to one's individual cognitive life, one selects as noemas whatever one finds to be one's fundamental cognitive commitments. These commitments are treated as "languageless" perceptions. The Diagram is applied as a function of those perceptions. It is required that all derivative cognitive commitments be connected by means of denotational connections, i.e. cognitions rising from the initial noemas are guided principally by the *law of sufficient reason*. The Diagram thus serves as a means for testing the grounding of one's fundamental ideas. For the most part, this *initial application* of the Diagram acts as a technique for emptying one's mind of its commitments, since the denotational connections between one's associated thoughts can rarely be provided on account of a lack of sufficient reason. The horizontal axis along which the noemas appear may be interpreted as a version of Brouwer's horizontal axis—namely, as the concrete procession of temporal moments conceived of as empty twoities. The emptying out of the vertical dimension of the diagram (due to the failure to discover denotational connections) will then appear as a process of continually returning from the incoherence of one's mental life to the empty substratum of the concrete stream of one's temporal existence—a return which can be experienced as facilitating the subjective exploration of one's authentic inner life.

The second aesthetic preference mentioned above is fulfilled by the consideration that the application of the theory of the Creative Subject to certain philosophical texts problematizes those texts, in the precise sense that every fundamental statement presented by such a text will ultimately be analyzed as some set of noemas, while every line of thinking derived from that set will become subjected to the law of sufficient reason. If the set of noemas is empty, the meaning of any derivative statement will be empty. If the set of noemas is non-empty, its members will form the foundation of the statement's meaning. As a matter of fact, Hennix is able to produce, by means of the application of The Theory of the Creative Subject, solutions to several outstanding problems such as the cosmological *Night Sky Paradox* (why is the night sky dark?) and the metamathematical *Consistency Problem of Arithmetic* (why is $0=1$ not an arithmetical theorem?) These solutions are possible because, in a manner analogous to the way Brouwer's twoities are capable of generating arithmetical series (or even the continuum of real numbers) the chain of noemas can be interpreted as a generalization of the generation of number series in general, so that number series of different lengths and distinct closure properties can be constructed or generated for different uses. Since the unveiling of the fact that different number concepts are frequently applied indiscriminately, as if they were the same number concept, the flexibility of the Theory of the Creative Subject and its models provides a cogent tool for various analytical tasks demanded by the deepest problems of philosophy.

Again, just as Brouwer's First Act of Intuitionism founds both culturally-relevant mathematical constructions and spiritually-motivated private worlds, so the models for the Creative Subject allow for a reconciliation between private enlightenment and public problematic.

A complete exposition of Hennix's thought is impossible in this introduction. I have therefore selected, somewhat arbitrarily, two topics for discussion which I hope may prove helpful in orienting the reader to *The Yellow Book*. The topics are : *Motivations for Post-*

Introduction

Wittgensteinian Formal Philosophy; From a Theory of Meaning to a Theory of Self-Illumination.

Motivations for Post-Wittgensteinian Formal Philosophy

The Yellow Book contains an interpretation/translation of certain key passages from Wittgenstein's *Tractatus*. I therefore wish to present a discussion of Hennix's concern with "formal philosophy"—a concern which may seem incongruous in what we might call the post-Wittgensteinian age.

It is well-known that Wittgenstein, beginning with his return to philosophy in 1929 (after hearing Brouwer lecture in Vienna), progressively abandoned the formal aspects of the enterprise undertaken in the *Tractatus Logico-Philosophicus*. This process of abandonment culminated, perhaps, in the powerful passages of *Philosophical Investigations* wherein Wittgenstein renounces many of the aims of formal philosophy: he renounces the ambition to purge natural language of its ambiguities and metaphysical obscurities; he renounces the hope of establishing, once-and-for-all, the logical structure underpinning language, material reality, and human thought; he renounces the pursuit of the chimerical and fascinating aspects of formal, logical thought— the appearance of depth that logical enigmas display, the appearance of adamantine, crystalline fixedness that logic promises to depict.

Given what I take to be Hennix's acceptance of the late Wittgenstein's renunciations, it is necessary to make clear what motivates Hennix's *use* of formal vocabulary and logical apparatus, and what justifies his apparent valorization of the *Tractatus*. (It may be remarked at the outset that in fact the later Wittgenstein by no means abandoned formal philosophy—see, for instance, *Philosophical Grammar*.)

In order to approach Hennix's conception of formal philosophy, it is necessary first to look at several possible post-Wittgensteinian motivations for the development of formal philosophies—motivations which certainly do *not* underlie Hennix's concerns.

Accepting Wittgenstein's reasons for renouncing the project of discovering a unique and universal formal philosophical system, one might still wish to make use of formal thinking as a sort of personal "therapy." Disciplining one's language, defining terms and issues, and working out systems of inference might appeal as a reasonable procedure for making one's way through the morass of intellectual life and the chaos of ordinary language. A private individual might choose to formalize his/her thought without making any claim upon public consensus and having no ambition to compete with other systems for universal truth.

Though not at all precluded, (during the Rhinebeck seminars Hennix often talked about the Socratic sense of philosophy as a kind of therapy for curing the confusions of the ordinary mind, and Wittgenstein himself often spoke of the work of philosophy being a kind of therapy aimed at eliminating conceptual maladies), such uses of formal philosophy are betrayed by the rhetorical attitude of Hennix's texts which seem at any rate to aim at a tone of ontological assurance that would make no sense at all for purely private utterance: Hennix's pronouncements are punctuated by moments that rise to an almost oratorical state of publicity, yet which at the same time limit their absolute claims by opening large questions about the very existence of the texts in which they are enunciated. Hennix states flatly that "reality" is a result of authority. But ontological authority is confined to conditions of textuality, since "Being" remains only formalizable as a term in a text, while the ontological status of the text itself, being unformalizable, is undetermined or incomplete. "Being" is also identified with "discernings" of boundaries occurring within "spaces" which are expressions of modalities of consciousness and guided/directed with the aid of collections of textual signs. "Thought" is inseparable from acts of attention and attention is directed by acts of thought, represented, say, by the shadows of the objects in the ontology of algebraic geometry, again understood as textual structures. But textual structures are inseparable from human aims and situations, so that most of the ontologies in the environment É documented in *The Yellow Book* are upon reflection both thoroughly constructivist and intuitionistic in that they are not asserted independently of their conditions of construction and are at every point dependent upon subjective acts of thought. They intentionally fail to guarantee the unique logical cosmos to which the rhetorical gestures in which they are developed suggest they might provide access.

The rhetorical appeals to what may appear to be a "vatic," oracular or authoritarian logical universality are never what the texts aim at conceptually. This actually should be obvious since Hennix operates throughout within Yessenin-Volpin's ethical position which at every point bestows upon the reader maximum freedom to reject even *proven* propositions: proofs are intelligible only in terms of situations where aims are defined; however, the apparent epistemological anarchy is subjected to the restriction that not all propositions are asserted and not all procedures recommended, because they are invariably *subject to individual choices and purposes*. (cf. Cage, Fahlström, Johns: The very *choice* suffices to make them one's own.) Proofs are not obedient to absolute principles but are defined as "fair means of establishing incontestable propositions convincingly,"—convincingness being a matter of the logic of individual acceptance. *Correctness* is relative to aims. Logic itself is defined as *the science of avoiding errors* and is demanded only where errors are 1. definable and 2. intolerable.

The requirement that logic specify principles of correctness regarding procedures for attaining aims where error is intolerable makes this format of logic not only useful, but an ultimate court of appeal for technological theories. And, of course, the use of formal theories within specific sciences is in Wittgenstein precisely the appropriate place for them. What is interdicted is the metaphysical extension of theoretical truths beyond the concrete sphere of their application. Something like Yessenin-Volpin's approach is in practice appealed to in every technological situation where other formal theories fail to accommodate themselves to practical exigencies, and, as developed by Hennix, Yessenin-Volpin may be seen as the true inheritor of Wittgenstein's late work.

But, again, the refinements and improvements of the logic of technology are not the principal aim of Hennix's work. (One might notice that Yessenin-Volpin's "eleutheric" ethics not only specifies the logic for technological projects, but also spells out and attempts to justify a hierarchy of ethical aims that would limit the probability of undesirable side-effects attendant upon the accomplishment of technological aims. Thus, the very theory that shows how technological projects might be facilitated actually establishes severe limitations upon their pursuit. Not all sufficient means are justified by an aim, although, necessarily, all *necessary* such means are.)

There is one more use that is commonly found for formal theories these days—though I don't think that the reader is likely to mistake this use for Hennix's own. Formalizations of psychological, social, aesthetic, and many other situations are frequently attempted where these situations are assumed to be understood intuitively well enough so that "common sense" can provide a criterion guiding the formalization. Such formalizations "explain" nothing that is not already contained in common sense, and, though "harmless" in the sense that they seem to pretend little regarding metaphysical universes beyond the familiar world, they can be quite pernicious in that they tend to reify and render fixed precisely the unquestioned assumptions of the social epoch under which they are undertaken).

What then can be said about Hennix's use of formal philosophy, given his disengagement from the above motives? To answer this question we pass to our second topic.

From a Theory of Meaning to a Theory of Self-Illumination

At a fundamental level of analysis, Hennix conceives the meaning of a proposition to be one of the main subjects of ethics (since the "reality" of meaning is also a function of authority), and, insofar as uncertainty of the meaning-relation is intolerable, the notion of proof will interfere with the notions of semantics. Thus, for Hennix, the meaning of a proposition is neither the concatenation of the meanings of its terms, the truth or falsity of such concatenations when measured against objective states of affairs, nor again the use to which the proposition is put in ordinary life. The logical meaning of a proposition is a function of the associated propositions which establish it.

This may seem to burden the propositions of everyday life to such an extent that they would for the most part become meaningless. This consequence is accepted by Hennix and points to an important second role that proofs play in Hennix's thought (beyond grounding propositions) which is, in fact, Hennix's main motive for doing formal philosophy: to allow the thinking consciousness to grasp the extreme degree to which its own perceptions and cognitions are vitiated with uncertainty. We live in a world saturated with cognitions for which we have no demonstrations, and to the degree that we allow this situation to go unexamined, we live in a world that is meaningless, when grasped from a certain ethical viewpoint.

This rather severe picture of the state of our cognitive universe, however, is not asserted as an authoritarian dogma (in the style of the Positivism of the thirties) to be imposed by a logical elite upon an ignorant populace; it is solely presented as an experience which can be turned into an instrument to be developed by those with sufficient experience to find a use for it. Because the concept of proof that Hennix adopts comes under the requirements of Yessenin-Volpin's "eleutheric" ethics, what counts as an acceptable proof for any proposition depends upon the aims inherent in the concrete situation of which the proposition is a part. A proposition is not true or false independently of its use context—yet, with formalization, this context is specifiable within parameters defined as narrowly as one requires. Thus, if one's concerns *were* in fact the trustworthiness of perceptions and cognitions in the everyday world, the proof-theory of meaning would be a particularly powerful tool for *displaying the extreme degree of unreliability inherent in our ordinary mundane beliefs*: one might thus be moved to examine the bases upon which the most elementary of our sensations, perceptions and assertions are constructed, and thus to find how frequently such constructions are, in fact, vacuous. The collection of meaningful assertions about the world turns out in general to be empty. Yet, the world, evacuated by the plethora of meaningless assumptions, cognitions, and assertions which normally cloud and clutter it, does not by any means vanish from view: in fact the empty space—the space vacated by the meaningless propositions—turns out to have a populous interior: an interior surveyable without interference from propositional assertions of any kind. (It is in this context that Hennix has formulated a particularly striking interpretation of Parmenides which to some extent also has been brought forward by S. Austin in the work referred to above, and which will be developed in the supplement to this issue of *Io*.)

I will conclude these remarks with a look at the specific sense in which the notion of "enlightenment" or "self-illumination" is developed in Hennix's work. I have, on the authority of the text itself, mentioned analogies with Zen meditation, and *The Yellow Book* is indeed organized around a series of "Model Subjects," a term which translates "Koan" and which Hennix borrows from the Shaw translation of *The Blue Cliff Records (Hekigan Roku)*. But the analogy is, to some extent, misleading, in that the disciplines or concatenation of disciplines, or forms of mental activity described by Hennix are not dependent upon a reference to Buddhist metaphysics as traditional or textual authority, nor upon participation in Buddhist institutions for authorizations of practice. (The parallels, however, between Hennix's soteriological epistemology and that of the Indian Logician Dharmakirti and the development of his thought within the Madhyamika Prasangika school go beyond those drawn in our introduction. I will certainly take this topic up in some detail in the future.) Though the enlightenment sought suggests Buddhistic forms of illumination, the methods and sociology, as it were, remain western, in the sense that they are conducted on the ground of the individual intellect and by the individual—responsible to his/her own intellectual and ethical life alone.

The substance of enlightenment or self-illumination for Hennix begins with an ascesis that aims at the establishment of the Creative Subject's independence from "mundane" forms of thought and from the ordinary language which saturates and controls ordinary cognitive life. One attempts with unremitting assiduity to identify the forms of thought which rule one's conscious and unconscious cognitivity, (the ultimate such forms are one's "noemas"), and through the application of analytical techniques one withdraws and reduces the streaming of associative thought to which they give rise.

But for Hennix it is insufficient simply to identify and suppress these noemas: the drift of linguistically saturated and alienated consciousness cannot be neutralized by simply observing the concrete content and activity of the movement of one's mind, for the general structures which determine this drift will remain operative unless they are pacified by constructions aimed precisely at consuming, replacing or reducing them. The general structural form of their production must be discerned. Such discernings will provide the preconditions for the reduction down to and ultimately of the noemas. Reduction of the noemas is possible, I repeat, through the comprehension of the general structure of thought and the assiduous application of this comprehension to one's own cognitive activity. Now, the discernment of the general form of thought involves an analysis of the linguistic forms in which thought is ordinarily encountered, and in this way Hennix's work has its roots in a tradition of poetics beginning perhaps with Hölderlin and extending through Mallarmé to Blanchot, Lacan, and Foucault. These writers delineate with great subtlety the problematic of the Cartesian subject with regard to its alienation and annihilation through the evolution of an awareness of language as such: the subject loses itself in its reproduction as the "I-speak" of a beginningless and interminable verbal stream, whose origin is inaccessible but whose existence as an ungraspable totality prefigures the entire of the space of individual linguistic consciousness.

But whereas the latter three of these writers base their linguistic thinking on the tradition of Saussure, Hennix works with concepts found in Montague Grammar that are in radical competition with specifically the Chomsky school. It is Montague that is, in any case, important for Hennix's work—in particular for his theories of abstract grammars and higher-order programming languages (logic programs). Montague worked from the premise that all language can be treated as a species of abstract mathematical structure, with the grammars of

Introduction

concrete languages falling out as special cases of purely formal possibilities constructed a priori from set-theoretical considerations. Hennix "lifts" Montague grammar and locates it within category theory, and thus brings it into contact with fundamental logical structures in an atmosphere where the possibility of freely choosing the basic logical laws underlying the linguistic structures becomes available for individually-determined conceptual use. By working with the "forgetful functors" "pullbacks" and the "empty category" in ways consonant with but not intended by the originators of category theory, Hennix, prescribes a method for eliminating cognitive commitments at a single stroke. Since the possibilities for this elimination are capable of apriori development, criteria can be envisaged by which the individual can test his/her advancement along the path. The application of such criteria are what make it possible for the individual to travel this path without guidance of tradition or teacher.

The condition of enlightenment aimed at, then, is a condition where the Creative Subject has rigorously liberated him/herself from the drift of ordinary cognitivity. The interest of such a condition actually *begins* at this point: one wants to imagine how the entire domain of human activity might be re-visioned and the entire domain of technology and culture revitalized under such conditions. Unfortunately, all that has been accomplished thus far is the determination of the "way", though it should be obvious that this has been a formidable task; nonetheless, it is only the beginning of what appears to be an inexhaustible journey. Though the path of *The Yellow Book* is thoroughly destructive (as is the initiation of Intuitionism through Brouwer's "first act") the consequence of this destruction is to introduce profoundly hopeful perspectives into a cultural and historical milieu for the most part dominated by intellectual triviality, myopia and fatigue.

Appendix I

The Two Basic Acts of Intuitionism

(from L.E.J. Brouwer's *Historical Principles and Methods of Intuitionism* (collected works)

......*The intervention of intuitionism* of which the first seems necessarily to lead to destructive and sterilizing consequences, whereas the second yields ample possibilities for recovery and new developments......

FIRST ACT OF INTUITIONISM
    completely separates mathematics from mathematical language, in particular from the phenomena of language which are described by theoretical logic. It recognizes that mathematics is a languageless activity of the mind having its origin in the basic phenomenon of perception of a *move of time*, which is the falling apart of a life moment into two distinct things, one of which gives way to the other, but is retained by memory. If the two-ity thus born is divested of all quality, there remains the common substratum of all two-ities, the mental creation of the *empty two-ity*. This empty two-ity and the two unities of which it is composed, constitute the *basic mathematical systems*. And the basic operation of mathematical construction is the *mental creation of the two-ity of two mathematical systems previously acquired*, and the consideration of this two-ity as a new mathematical system.
    It is introspectively realized how this basic operation, continually displaying *unaltered* retention by memory, successively generates each natural number, the infinitely proceeding sequence of natural numbers, arbitrary finite sequences and infinitely proceeding sequences of mathematical systems previously acquired, finally a continually extending stock of mathematical systems corresponding to "separable" systems of classical mathematics.

THE SECOND ACT OF INTUITIONISM
    which recognizes the possibility of generating new mathematical entities:

Introduction

firstly in the form of *infinitely proceeding sequences* $p_1\ p_2\ldots$, whose terms are *chosen more or less freely from mathematical entities previously acquired;* in such a way that the freedom of choice existing perhaps for the first element p_1 may be subjected to a lasting restriction at some following p_ν, and again and again to sharper lasting restrictions or even abolition at further subsequent p_ν's, while all these restriction interventions, as well as the choices of the p_ν's themselves, at any stage may be made to depend on possible future mathematical experiences of the creating subject;

secondly in the form of mathematical species, i.e. *properties supposable for mathematical entities previously acquired,* and satisfying the condition that, if they hold for a certain mathematical entity, they also hold for all mathematical entities which have been defined to be equal to it, relations of equality having to be symmetric, reflexive and transitive; mathematical entities previously acquired for which the property holds are called *elements* of the species.

With regard to this definition of species we have to remark firstly that, during the development of intuitionist mathematics, some species will have to be considered as being re-defined time and again in the same way, secondly that a species can very well be an element of another species, but never an element of itself.

Two mathematical entities are called *different* if their equality has been proved to be absurd.

Two infinitely proceeding sequences of mathematical entities $a_1, a_2, \ldots$, and $b_1, b_2, \ldots$, are called *equal* or *identical* if $a_\nu = b_\nu$ for each ν, and *distinct,* if a natural number s, can be indicated such that a_s and b_s are different.

Fran Lej Brouwer Collected Works I
Ed A Heytin North-Holland Publishing Company, 1975

Appendix II

Excerpts from
Life, Art and Mysticism
by L.E.J. Brouwer   (1905)   (Collected Works)

Originally man lived in isolation; with the support of nature every individual tried to maintain his equilibrium between sinful temptations. This filled the whole of his life, there was no room for interest in others nor worry about the future; as a result labour did not exist, nor did sorrow, hate, fear, or lust. But man was not content, he began to search for power over others and for certainty about the future. In this way the balance was broken, labour became more and more and more diabolical. In the end everyone wielded power and suffered suppression at the same time. The old instinct of separation and isolation has survived only in the form of pale envy and jealousy.

TURNING INTO THE SELF

Having contemplated the sadness of this world, look into yourself. In you there is a consciousness, a consciousness which continually changes its content. Are you master of these changes? You say no, for you find yourself placed in a world which you have not created for yourself, you are bewildered by its continuous state of flux.

The content of your consciousness, however, is to a great extent determined by your moods and these are within your power. Is the motto "Control your passions" only an empty phrase to you? Sometimes you must have experienced that religious feeling of escape from your passions, from fear and desire, from time and space, from the whole world of perception. Finally, you do know that very meaningful phrase, "turn into yourself." There seems to be a kind of attention which centres round yourself and which to some extent is within your power. What this Self is we cannot further say; speaking and reasoning is an attention at a great distance from the Self; we cannot even get near it by reasoning or by words, but only by "turning into the Self" as it is given to us.

Introduction

This "turning into oneself" is accompanied by a feeling of effort, it seems as if some inertia has to be surmounted, that your attention is strongly inclined to linger where it is, and that the resistance felt in the move towards the inner Self is much greater than in the move away from it. If, however, you succeed in overcoming all inertia, passions will be silenced, you will feel dead to the old world of perception, to time and space and to all other forms of plurality. Your eyes, no longer blindfolded, will open to a joyful quiescence.

. . . .

Now you will recognize your Free Will, in so far as it is free to withdraw from the world of causality and then to remain free, only then obtaining a definite Direction which it will follow freely, reversibly. The phenomena succeed each other in time, bound by causality because your coloured view wants this regularity, but right through the walls of causality "miracles" glide and flow continually, visible only to the free, the enlightened.

MAN'S FALL CAUSED BY THE INTELLECT

Without pain the free man sees mankind cast down by fear and desire, by avarice and lust for power, by time and space, aimlessly wandering without wings, incapable of lifting itself in self-reflection chained to the spawn of time and space, the Intellect, which has become fossilized in the form of the human head, the symbol of man's fall.

. . . .

This highly esteemed Intellect has enabled and has forced man to go on living in Desire and Fear, rather than—from a salutary sense of bewilderment—take refuge in self-reflection. Intellect has made him forfeit the staggering independence and directness of each of his rambling images by connecting them with each other rather than with the Self. In this way the Intellect made him persevere in a state of apparent security in a "reality" which man in his arrogance had made himself, which he had tied to causality, but in which eventually he must feel totally powerless.

In this life of lust and desire the Intellect renders man the diabolical service of connecting the two images of the imagination as means and end. Once in the grip of desire for one thing he is made by the Intellect to strive after another as a means to obtain the former.

....

The act aimed at the means always overshoots the mark to some extent; the means has a direction of its own, at an angle, however small, from the end. It acts not only in the direction of the end, but also in other dimensions. Man's blinkered view prevents him from recognizing the sometimes very detrimental effects of such action, but worse the end is gradually lost sight of and only the means remains.

In this sad world, where a clear view of all human activity is impossible, a world dominated by Drill and Imitation, the other offspring of Fear and Desire, many recognize as an end what was originally only a means. They seek what we might call an end of second order and in so doing may discover a means again out of line with the corresponding end. If this deceptive jump from ends to means is repeated several times, it may happen that a direction is pursued which not only deviates into other dimensions but even opposes the direction of the original end and therefore counteracts it.

Industry originally supplied its products in order to create the most favorable conditions for human life. But one ignored the fact that in manufacturing these products from the resources of nature one interfered with and disturbed the balance of nature and of human conditions, thereby causing damage greater than the advantages of these products could ever justify.

But worse: manufacturing these industrial goods has become an end in itself, new industries were called into existence merely to supply instruments to facilitate production. Another blow was dealt to the balance of nature. Raw materials were recklessly seized from far away lands, commercial and naval enterprises were created with all their physical and moral misery, all leading to oppression of one people by another.

Now that the Self had been abandoned, the Self that knows all about the past and the future, man grew anxious about the future and craved for power to predict. Science originates in this desire to predict, in its early stages it is completely subservient to industry. Science asserts generalizing propositions in and about the world of perception; these will come true as long as it pleases God, sometimes they are contradicted by the facts. The scientists then exclaim: "Yes indeed, but we had made this or that tacit assumption." In their incompetence they then set

Introduction

about complicating the proposition further and making so called improvements.

Science does not remain confined to serving industry, again the means becomes an end in itself and science is practised for its own sake. A further aberration has been the concentration of all bodily awareness on the human head thereby excluding and ignoring the rest of the body. At the same time man became convinced of his own existence as an individual and that of a separate and independent world of perception. At that stage the full extent of the deviation of human scientific thinking became clear, for scientific thinking is nothing but a fixation of the will within the confines of the human head, a scientific truth no more than an infatuation of desire restricted to the human mind. Every branch of science as it proceeds, will therefore always run into trouble; when it climbs too high it becomes blindfolded in even more restricted isolation, the remembered results of that science take on an independent existence.

The "foundations" of this branch of science are then investigated and this soon becomes a new branch of science. Then one begins to search for the foundations of science in general and knocks up a "theory of knowledge." As they climb higher and higher trouble increases and in the end everyone is thoroughly confused. Some in the end quietly give up. Having thought for a long time about the elusive link between the intuiting consciousness—which itself develops from the world of phenomena—and this world of phenomena (which again itself exists only through and in the form of the intuiting consciousness)—a confusion which originated in a sinful foundation of the world of intuition, they then plug the whole with the concept of the Ego which was self-created with and at the same time as the phenomenal world. And then they say, "Yes, of course, something must remain incomprehensible and that something is the Ego that comprehends." But there are others who do not know when to stop, who keep on and on until they go mad; they grow bald, short-sighted and fat, their stomachs stop working properly, and moaning with asthma and gastric trouble, they fancy that in this way equilibrium is within reach and almost reached. So much for science, the last flower and ossification of culture.

Even if Conscience penetrates into the enclaved categories, man's attention is diverted away from it by the strongly felt stimulation and satisfaction of other needs, or it is assimilated, i.e. it is recognized as a need within the closed system and capable of satisfaction in the system. The main function of art and poetry but also of religion is to silence the human conscience by recognizing this need and by apparent but not real satisfaction. Art and religion in this world are only morphine industries, the yearning for a better life is only lulled into sleep or into a state of torpidity.

RECONCILIATION

This corrupt world, as you now recognize, only exists because of its very corruption, its deviation from the paths of rectitude. A world of righteousness seems to you as contradictory as your own mortality.

In language transcendent truth—even less than immanent truth—cannot be revealed without causing an outrage. A clear and true statement, seriously and emphatically pronounced, is no more acceptable than the manifest performance of a miracles.

. . . .

In language transcendent truth seems to be the prerogative of imitators who have vaguely understood the words of the prophet and recognized them as truth, but only after watering it down to suit themselves.

. . . .

They will liberate the world of all sorts of vices, stupidity and injustice, and be hailed as the benefactors of mankind but leave mankind as miserable as it was.

. . . .

The advice to get rid of the intellect, "this gift of the devil" is qualified by some added remark in defence of the viewpoint of the repudiated intellect, such as for example: "The structure of nature is so infinitely subtle and complex that your intellect will never fully grasp it and so you will never find there the stability you aim for." For those who relinquish the intellect, however, the world is anything but subtle or complex: it is immediately clear: it appears subtle only to the intellect that struggles laboriously and sees no end to its struggle.

. . . .

Look at this world, full of wretched people, who imagine that they have possessions, afraid they might lose them, always hopefully toiling in an

Introduction

effort to acquire more; look at people who strive after luxury and wealth, at those whose riches are secured, whose stocks and shares are safely deposited, but who nurture an insatiable appetite for knowledge, power, health, glory and pleasure.

Only he who recognizes that he has nothing, that he cannot possess anything, that absolute certainty is unattainable, who completely resigns himself and sacrifices all, who does not know anything, does not want anything and does not want to know anything, who abandons and neglects everything, he will receive all; to him the world of freedom opens, the world of painless contemplation and of —nothing.

Fran Lej Brouwer Collected Works I
Ed. A Heyting North-Holland Publishing Company, 1975

Symbolic Nature

Don Byrd

from
The Poetics of Common Knowledge

> ... modern European philosophy, even before the ego cogito but certainly from then on, situated all men and all cultures—and with them their women and children—within its own boundaries as manipulable tools, instruments. Ontology understood them as interpretable beings, as known ideas, as mediations or internal possibilities within the horizon of the comprehension of Being.
> —Enrique Dussel

1. The Two Cultures

The Pursuit of the Infinite

The complaint of the seventeenth-century modernists against the scholastic tradition was justified: scholasticism neither progressed nor was it interested in progression. Rather it circled a persistent core of problems, inquiring about the nature of God, society, and man, mediating between the universal categories of logic and unique actualities. The static images which were produced in this contemplation bore witness to a fundamental incommensurability and, thus, to a perpetual ontological crisis. The maintenance of a just relationship between *eidos* and *nous* was a massive contemplative chore. In the Aristotelian scheme each object is unique and each event requires an unique explanation. When a stone falls, according to traditional physics, it suffers an inward transformation which carries it to its "natural place." Until something happens there is nothing to explain; until the stone falls it *is* in its natural place. This knowledge yields itself case by case; the disposition of each entity must be determined.

The seventeenth-century search for a "method" of knowledge was motivated by the hope of discovering entire classes of phenomena which corresponded to the mathematical species of algebra. Unlike classical arithmetic, the new mathematics developed by Vieta and Descartes did not require an actual counting of actual things. Vieta speaks of a finding of finding or "the zetetic art [which] does not employ its logic on numbers—which was the tediousness of the ancient analysts—but uses its logic through a logistic which in a new way has to do with species." Mathematical analysis deals not with particular quantities but classes of quantities, mathematical potentialities or functions (e.g., $y+f(x)$), which describe forms of relationships between *possible* individuals. We know, for example, the *form* of the acceleration of a ball on an incline plane; it can be expressed as an algebraic formula. The formula itself tells us nothing, however, about particular balls and particular inclined planes; the relevant quantities must be measured and substituted for the generalized expressions of the formula. The functions in terms of which classical physics is expressed are independent of any *particular* world, and their advantage is precisely their avoidance of particular and sensuous means of expression. Unlike classical *mimesis,* seventeenth-century methodology modeled a world of which there is no unique image; it defines not images but *infinite classes* of images. Our world, the "real" world, is a contingent instance of the *possibilities* described by the formulae.

If Euclidean geometry is timeless, Cartesian geometry is also paradoxically spaceless. It is removed from the realm of concrete intuition altogether. From the time Vieta's *"ars magna"* found a "place in the system of knowledge in general," Jacob Klein writes, "... the fundamental *ontological* science of the ancients is replaced by a *symbolic* discipline whose ontological presuppositions are left unclarified." Thenceforth, the being of all things was not number in the classical sense but an unlocated, formal property. It could be variously identified with ego, nature, or God, and it could be quickly and conveniently moved from one of these sites to another like the pea in the shell game. The modern view has a built-in fudge factor. The countable and accountable world was replaced with a formal possibility, the content of which is arbitrary and, from a theoretical point of view, even superfluous.

Herman Weyl speaks of the seventeenth-century "leap into the beyond." Thereafter, the efficacious cultural forms were unavailable to intuitive scrutiny. The world to which the senses are responsible leaked out into the infinitely large and small. In an attempt to clarify the attitude of the algebraists, Weyl alludes to certain Indian and Buddhist texts which revel "in the possibilities of producing and designating prodigious numbers by means of the positional system, that is, by combination of addition, multiplication, and exponentiation." He adds: "In spite of their fantastic aspect there is something truly great about these efforts; the human mind for the first time senses its full power to fly, through the use of symbol, beyond the boundaries of what is attainable by intuition." One feels this excitement in the writings of the seventeenth-century mathematicians. Vieta's *Introduction to the Analytic Art* (1591) ends by appropriating "to itself by right the proud problem of problems which is: TO LEAVE NO PROBLEM UNSOLVED." And even as he lays out the rules of his methodology which insist upon the need for caution, Descartes notes simply, "there is no need for minds to be confined at all within limits." The question of the ontological status of the universal was swept away by this enthusiasm. Herman Weyl:

> The leap into the beyond occurs when the sequence of numbers that is never complete but remains open toward the infinite is made into a closed aggregate of objects existing in themselves. Giving the numbers the status of ideal objects becomes dangerous only when this is done. The belief in the absolute is deeply implanted in our breast; no wonder, then, that mathematics was bold and naive enough to perform the leap. Whoever accepts as meaningful the definition 'n' is an even or odd number according as a number x does or does not *exist* such that $n=2x$, which refers to the infinite totality of all of numbers ... , already stands on the other shore; for him the system of numbers has become a realm of absolute existences which is 'not of this world' and from which only gleams here and there are caught and reflected in our consciousness.

This position is often called "Platonic", and there are Platonic arguments which indicate a clear longing for the other shore. In fact, however, Plato knew only a worldly mathematics in which the sequence of numbers, open toward the infinite and the infinitesimal, is never "made into a closed aggregate existing in themselves." Without entering the fray over the Platonic conception of number, which is fought by scholars of formidable learning and dialectical skill, it can be safely said that the modern number, unlike the ancient, does not depend upon actual counting and actual countable entities. The modern view is not simply *more* abstract but a different order of abstraction. This is the first appearance of what Hegel will call "the right infinity," infinity as a real quantity, not just an endless succession. It is the quintessential modern concept, the concept which separates us from the Greeks. It involves logical absurdities which cannot be resolved, but they are absurdities which now are so embedded in the fabric of life that they are nearly imperceptible.

The complexities of this situation created new a psychology: like an algebraic expression, the Cartesian ego was an ideal object which could only be articulated in formal language, especially the formal language of poetry and art. The rise of mathematical analysis and the Miltonic tradition in poetry were cognate developments. The new poetry represented an ideal self as mathematics represented an ideal world, both of which are infinite and thus in some sense incomprehensible. The rhythm of the accentual-syllabic line, unlike classical verse, was not based on a real counting of intuitable quantities; it was rather a relationship, like an algebraic function, which defined rhythmic species and might be realized by a range of possible performances.

In the new regime each individual soul became the site of "paradise within," as the archangel Michael tells Adam at the end of *Paradise Lost* and, at the same time, the cultivator of the garden which reveals God's will by allegory. There were awful paradoxes hidden in this formula, but they were obscured by the substitution of the new methodology for the slower and less powerful techniques of *mimesis*. The dissociation of sensibility which T.S. Eliot notes in Milton was not a failure of art but a necessary feature of the new program. In a revealing book, *Milton and the Postmodern,* Herman Rapaport argues poststructurally that "Milton is not a poet of the logocentric, the theocentric book, but a poet who banishes the signified." As we will see, every poet who absorbed and continued the methodological tradition from the beginning both does and does not banish the signified. In one sense, the poets obviously banish it: they create fictions. At the same time, however, they create special fictions which are truer than the truth of the signified. The mere signified exists in an inferior world, a world itself banished, and substituted for. For the solipsist, which Milton was, the place of the signified is taken by more signifiers. The poststructuralists, in this sense, hardly provide a critique of the tradition: they merely expose some of the tricks which it has made effective.

The solipsism was rigorous. Never before had individuals felt so keenly the responsibility for their own beliefs. For the few the intellectual challenge was unsurpassed; for the many only the enthusiasm of hymn-singing and Bible-thumping made the interior garden seem real at all. In time the intensities and insecurities of the situation became unbearable, and, after the great wave of religious psychosis and strife past, the rational spirit of the theology was attached to science, and the enthusiasm was attached to art. These two modes of the aesthetics of self came to pass for the objective and subjective realms, the one expressed by science, the other by *sentimental* reference to the tradition. Humans, thus, became subject to their own self-creations, metaphors in which they were weak and unreliable but oddly empowered as agents of change because the Other was created in the image of their reason. They were simultaneously master and slave, god-like creators of replicas on which they modeled themselves, disciplining themselves to mechanism. Although idealism and materialism were their bleak *philosophic* alternatives, philosophy itself was largely irrelevant, except as symptom or official policy. A new psychology had taken hold, and inside it there were no true contraries as Blake would note. Maintaining a subjectivity so large and commodious that physical mechanism and rational freedom could be simultaneously maintained was a massive cultural project.

Ideal Language

"Descartes" is our convenient name for the site of cognitive excitement which marks the essential assumption for the community of modern thinkers. The historical personage and his writings are by no means irrelevant, but the site as such is dependent upon neither. We honor him, rather than Vieta or Galileo and Bacon, who were prior to him in important respects, not only because he independently developed powerful techniques of mathematical analysis, but he also identified mathematical potentialities with pure space, thus, creating the context of modern physics and, most significantly, for our purposes, defined the relationships between the common language and formal languages in terms which have persisted as fundamental structures of the culture. Or, if we emphasize the subject, rather than the object, we can say that Descartes gave us the first complete image of the modern self, the solipsistic demiurge of modern technological psychology, an entity whose judgements call an unlocated and problematic world into being.

The Cartesian strategy creates a language of language, a writing of writing. It is evidence and understanding of this doubling that we seek in Descartes' work. It was the culmination of a profound culture change, closely associated with the development of printing and the development of linguistic forms appropriate to it. The question-and-answer format of classical dialectics gave way to a series of manic monologues. Descartes' *Rules for the Regulation of the Mind* consists not of questions and answers but propositions and commentaries. The modern era is initiated by *self* interpretation, not catechism or discussion. In addition to Descartes, one might think in this regard of the essays of Montaigne, Bacon, and Burton, or the solipsistic implications of the novel in the instigating hands of Cervantes and Defoe. The linearity of print-bred intelligence depends upon writing as an image of abstract temporality, a temporal sense which distinguishes it clearly from the linear duration of oral poetic performance, the temporal play of classical dialectics, and the oratorical time of classical rhetoric. The change from qualitative science to quantitative science paralleled a shift in emphasis from ontology to epistemology, from grammar to rhetoric, and from poetry to prose.

The seventeenth-century implementation of a mechanical information-handling technology required the revision of fundamental beliefs and institutions. The proposals for reform were diverse—some more or less useful; some, from our perspective, more or less absurd. Murray Cohen, in *Sensible Words: Linguistic Practice in England, 1640-1785,* gives a useful account of the linguistic projectors:

> ... the common topics include not only proposals for altering teaching of Latin, introducing instruction in English, inventing new languages, and correcting orthography, but also ideas for fixing pronunciation and assembling common words, phrases, technical words, proverbs, idioms, and archaic words. In addition, projectors tried developing the quickest shorthand, teaching the deaf and dumb, discovering the original language, promoting universal speech, organizing language mathematically, equipping the language for science, discovering the relationship between words and things. They also looked forward to providing English merchants with the advantage of a universal grammar, spreading the word of God, adapting education to language learning, communicating secretly or at a distance, interpreting Chinese characters and Egyptian hieroglyphics, and at last, repairing Babel.

A remarkable number of these projects, as diverse as they are, share a common theme of breaking the relationship between language and speech. The animadversions of the linguistic projectors on the centrality of the rhetorical voice anticipated and, in a sense, outdid the twentieth-century critique of voice led by Jacques Derrida. Respect for the voice as the guarantor of self-presence and for the associated metaphysics of the originating Word was undermined as far as the primary intellectual modes were concerned. In the seventeenth century, as in our own times, the fact that European writing was a mere transcription of vocal sounds which in turn referred to objects was a cause of special consternation. Phonetic language seemed to double the distance between a description and its object. As early as 1615, Matthew Ricci gave an influential account of Chinese ideographs, and proposals for visual languages were common. Even Francis Bacon commented on the advantages of glyphic over phonetic writing. Not only teaching the deaf but teaching as if *everyone* were deaf became a primary undertaking of a culture gearing up for book literacy and mathematical technology.

By the end of the century, the outlines of a scheme which could compensate for the shortcomings of phonetic writing was articulated. The Cartesian solution to the problem was to create two languages: one formal, descriptive, and geometric; the other expressive, decorative, and historical. Although complex and often inefficient cultural institutions were required to oversee the practical relationship between these primary linguistic zones the strategy afforded unprecedented intellectual leverage.

Descartes understood at least implicitly the necessary relationship between language reform and technology. In a letter to his friend Mersenne about the time he was writing the *Regulae,* he suggested that an ideal language, which, like mathematics, would reveal the truth by virtue of its structure, is possible: "Order is what is needed: all the thoughts which can come into the human mind must be arranged in an order *like the natural order of numbers."* The suggestion here is related to the *mathesis universalis* of the *Regulae* as well as Leibniz's plan for a universal calculus, Frege's symbolic logic, Russell's logical atomism, or any language which depends upon the identification and manipulation of "the simple ideas in the human imagination out of which all human thoughts are compounded." Descartes goes on to say: "The greatest advantage of such a language would be the assistance it would give to men's judgment, representing matters so clearly that it would be almost impossible to go wrong." The conception of a language which grammatically or structurally disallows falsity was a central philosophic fantasy from Descartes to the early Wittgenstein and it persists, at least as a motivating fantasy, among certain researchers in artificial intelligence.

Descartes warns Mersenne, however, that he should "not hope ever to see such a language in use. For that, the order of nature would have to change so that the world turned into a terrestrial paradise; and that is too much to suggest outside of fairyland." The recognition that such a language is possible but that it cannot be implemented without *a change in the order of nature* brings us to the moment of the technological crisis, the disjunction between what the order of nature produces of its own accord and what is possible.

Most people, perhaps all people most of the time, are hopelessly mired in unreason. If the structure of reality is determined logically and, therefore, unreasonable people are requirements of that structure, how does one establish the grounds for truth *outside* the habitual and determining language of the peasant or, for that matter, the intellectual salons? What conditions make reason possible? Although Descartes did not pose the question in quite these terms, which belong to Kant and the following century, he confronted the problem directly. If language is determined by habit and thought by language, how might it be possible to establish an arena of free intellectual renovation?

As an "instrument of knowledge," Descartes feels, the natural languages derive from or are secondary to a mathematical science which contains "the primary rudiments of human reason." At the same time, his inability to develop mathematics *without* the "derivative" language is clear; he must have it to state the rules. This is a logical problem of the first order, and Descartes makes the only possible arguments: 1) truth can be established only in an ideal language, and 2) the natural language *in the ironic and sentimental mode* can meet its metalinguistic requirements. He draws the analogy to the person who would practice the craft of a smith but who lacks the necessary tools and equipment. It would be necessary, he tells us, for the person first to fashion the required items in stone and wood and, then, in turn, to use these to make tools of more suitable but more intractable materials. "This method of ours," he writes, "resembles indeed those devices employed by the mechanical crafts, which do not need the aid of anything outside of them, but themselves supply the directions for making their own instruments." The ideal language is a subset of the natural language, and, if a few clearly defined terms can be established, the remainder of the language can be constructed from these nodes of clarity, or at least that is the hope.

It is not clear, however, that one linguistic mode is the iron to another linguistic mode's stone. If a language as inclusive and fluent as the natural language is developed, how can it avoid all of the complexity and imprecision of the natural language? Descartes was athwart the problem of ideal languages which the later Wittgenstein so cogently exposes: "When I talk about language (words, sentences, etc.) I must speak the language of every day. Is this language somehow too coarse and material for what we want to say? *Then how is another one to be constructed?*—And how strange that we should be able to do anything at all with the one we have!" As a logical problem, it is insoluble. It is possible, however, to create cultural institutions with enough tolerance that the problem can be absorbed, at least in sufficient measure to make ideal languages useful to technology, however uncertain their metaphysical foundations.

The Second House of Culture

The modern strategy for creating ideal languages has been to accept a traditional language as a kind of metaphysical halfway house, while articulating ideal structures which are alien to it. Before undertaking the process of doubt in *Discourse on Method,* Descartes notes that we do not rebuild the house in which we live without furnishing ourselves with "some other house in which we may live commodiously during the operations," so he adopts certain maxims to live by, as he carries out the reconstruction of philosophy—for example, "to obey the laws and customs of my country, adhering firmly to the faith in which by the grace of God, [he] had been educated ... " That is, he provisionally accepts the tradition which he is questioning and undermining.

The epistemic border which appeared in the seventeenth century was not between the known and the unknown but between two normative languages—one logical and one poetic, one in the new formalism and one a formalization of the natural language. Although the supposed gap between the two cultures has occasioned much impassioned graduation-speech rhetoric, the two languages and the two cultures which serviced them were both essential. Thus, Descartes, in *Regulae:*

> To study the writings of the ancients is right, because it is a great boon for us to be able to make use of the labours of so many men; and we should do so, both in order to discover what they have correctly made

out in previous ages, and also that we may inform ourselves as to what in various sciences is still left for investigation. But yet there is a great danger lest in a too absorbed study of these works we should become infected with their errors, guard against them as we may.

"The ancients" is a code term for what would be known as "culture" in the nineteenth century and "the humanities" in the twentieth. Descartes is thinking not only of the ancient texts but also the entire exegetical tradition, the lock, stock, and barrel of his own formal education, and, more generally, the tradition of letters itself. In counseling casual regard for the ancients, Descartes hits the keynote of subsequent rationalistic, technological, liberal thought. Since the seventeenth century, the *literary* tradition, in the broadest sense, the tradition of literature, of philosophy, to the extent that it has not been a pursuit of formal symbolic systems, of history, of colloquial thought in general, has been implicated in the sentimentality of a pre-philosophic Golden Age. The modern West, the culture which has taken modernity as its definitive quality, has obsessively fantasized about its relationship to the ancients. Never has a culture so poised itself upon the dream of another so distant in time and space. Like Chronos with a time machine, we have devoured our ancestors. In audacious and inspired acts of scholarly piracy, we have appropriated classical images which are sensuous, timeless, and wholly inappropriate to the modern impulse. The fantasy has not been hampered by excessive knowledge; we do not know much about the Greeks. It is, at any rate, indirectly the Greece of the scholars which has figured in the modern imagination, but from the Italian Neo-Platonists, Marlowe, and Shakespeare, to Joyce and Pound, the reservoir of classical imagery has been the target for a nervous meditation, a recalling and a casting away of a genealogy which we have created to justify and temper our rapaciousness.

The nostalgia for a time in which the particular is perfectly accommodated by the universal and vice-versa is the instigating sentimentality of modern rhetoric. Georg Lukács writes: "For the question which engenders the formal answers of the epic is: how can life become essence?" This is a clear and important formulation. In a pre-literate poetry a great deal is forgotten, but the parts which are retained all have equal status. No part of the Homeric text exercises control over the remainder. It is the most parsimonious use of memory, and that which is remembered is worn down to the narrative nubs of myth. There are no irrelevancies: each image and piece of information retains its integrity; the value of information is not absorbed into a structural hierarchy. In Homer, the essences are made manifest, not named. The problem of the individual is solved by the fact that the epic discourse is structured by proper nouns; it never occurs as a problem because the problem of the universal never arises.

The origin to which the modern West recurs is as much historical as metaphysical or, to say the least, the gap between the requirements of the theory and the practical workings of the culture is immense, and it is only the hopelessly intellectual who assume that the theory is more than an elegant afterthought. We conceive of a philosophic apocalypse in which thought will be sublated and the possibility of Odyssean action will reappear. Both of these conceptions, both the Golden Age and the dialectic which leads *forward* to it, are products of the philosophic tradition, and together they conspire to constitute a project, the program of which is nowhere clearly articulated.

As a rule of thumb in reading modern philosophy, one can assume that the gist of the position is never contained in an isolatable passage. The center of interest is the propulsion toward the coherent truth. To uncover the movement of the philosophic currents at a given time, therefore, it is necessary to examine the given details against the vastest background. The meanings of texts appear only in relation to a tacit knowledge of the complete tradition. This fact discourages scholarship. Although the academy creates the appearance of isolated subject matters for microscopic scrutiny of expert researchers, it is not possible to test the generalizing sweep of even so cautious a philosopher as Descartes—to say nothing of Hegel or Freud—against all of the relevant material. The histories which are implicit in modern thought, however, are not factual but mythological. Our dream-time happens also to have been an historical epoch, from which we can recover texts, artifacts, ruins, and so forth. Our arguments over the nature of Greek social organization, the translation of pre-Socratic philosophic fragments, or the development of vase painting, therefore, are in effect theological.

Unlike the humanist scholars, the scientists and mathematicians were not restoring ancient texts but the pure thought of the Golden Age of science which we know, if at all, in late and decadent form. Of this blest time, Simon Stevin, the mathematician, writes, "We call the wise age that in which men had a wonderful knowledge of science which we recognize without fail by certain signs, although without knowing who they were, or in what place or when." Descartes goes so far as to accuse the ancients of harboring the secret of their discoveries for mercenary reasons—a practice which was not uncommon among Descartes' contemporaries. The scientists and mathematicians were "rearticulating" a world which never existed. They projected their own epistemological obsessions on a tradition which was largely interested in ontology, and the contradictions which were inherent to their project were deep and generative. The Modern dialectic has been responsible to logic, to empirical evidence, and to personal freedom. The primary philosophic enterprise has been to discover a synthesis in which these contradictory claims can be satisfied. Such syntheses are, of course, inherently unstable: personal freedom finds itself contending with the demands of *both* of the other terms. It was in the creation of a theoretical world that a reasonable compromise seemed possible. The modern impulse has been to create a second world, which replicates the first, a universal techno-environment, including perhaps replications of living organisms and, at least in computers, human intelligence.

The Unacknowledged House

The two cultures, however, required a third. The languages of mathematics and poetry required the mediation of prose—a medium in which the ideal languages could be laundered. Jeffrey Kittay and Wlad Godzich write, "... prose is considered omnipresent ... Prose is meant to have no place; prose does not happen. Prose is what assigns place." The universal prose which appeared in the seventeenth century, was not so much a literary genre as a new world, almost nature itself, which appeared to stand between the objective idealism of science and subjective idealism of poetry. Kittay and Godzich go on to say:

> In relation to verse or indeed any other form, prose assumes the position of matter.... It is there from the beginning, as the *hyle* of the world, and it is what will remain after the destruction of whatever may have been imposed on it. Unlike ancient *hyle,* however, prose is not inert; it does not wait for the inspirational breath to set it in motion, it animates and motivates, disposes, arranges, assembles, and orders by itself. This is a position that prose has staked out for itself, or, if you will, that culture has demanded.

As a self-energizing and motivating medium, prose is both a cultural motor and a persistent source of potential anarchy at the very heart of the culture. It is the infinite resource of printed language: an undefined medium which implicitly encloses as a system all of the combinatorial possibilities of the alphabet. It encloses the infinite, thus creating an insoluble puzzle for the subject which identifies with its logical possibilities rather than its factual condition.

The terms of the debate over modern prose style were defined more than fifty years ago in essays by Morris W. Croll and R. F. Jones and have not been seriously disturbed by subsequent considerations. Croll emphasizes the reaction against the rhetorical style of Cicero by the Senecans, Lipsius, Montaigne, and Bacon, thus seeing the matter as a continuing adjustment of modern practice to classical models. Jones, on the other hand, emphasizes the relationship of prose to science. He shows that many of the important concerns were immediate and utilitarian rather than literary. Therefore, he argues that the thoroughly modern prose style did not develop before the last half of the seventeenth century. To say the least, there were two strains of prose in the seventeenth century, and the problem has revolved around how to characterize them. As Stanley Fish points out, the opposition can be seen in political, social, religious, or epistemological terms. He notes, "To the paired terms of my predecessors—Anglican-Puritan, Painted-Plain, Ciceronian-Senecan, Scientific-Rhetorical, Utilitarian-Frivolous —add a new pair, Self-Satisfying and Self-Consuming." Fish rightly shows that the universal acceptance of the plain style after the Restoration was a result not of the triumph of this or that party but of an epistemological shift which all of the parties more or less shared.

Prose style, thereafter, became the intellectual currency, the basis for a dialectic which was not possible as long as different styles represented radically-divergent logical spaces. Allied with the newly available abstraction, prose was an awesome tool, and it proliferated as the utility of public information and entertainment. Formal beauty and expressiveness were incidental values. It was the raw material of thought or experience. The *formalism* had been relocated. Despite the divergence of the sources from which they trace the development, Croll and Jones agree that prose is *natural* language. Of Thomas Browne, Croll says: "He writes like a philosophical scientist making notes of his observation as it occurs. We see his pen move and stop as he thinks. To write thus, and at the same time to create beauty of cadence in the phrases and rhythm in the design—and so Browne constantly does—is to achieve a triumph in what Montaigne called 'the art of being natural ....'" (This line of thought, given another half turn, justifies the automatic writing of the surrealists.) Jones, on the other hand, emphasizes the naturalness of the self-effacing style which was favored by Puritan preachers and the scientists of the Royal Society, which are by no means mutually-exclusive groups. John Wilkins, for example, who was one of the founders of the Royal Society, authored important statements on the style of both sermons and scientific reports. Jones finds the origins of modern prose style most clearly characterized in the often-quoted passage from Bishop Sprat's *History of the Royal Society* (1667). In keeping with the necessary strategy of linguistic renovation, Sprat appeals to the precedent of a Golden Age:

> They have ... been most rigorous in putting in execution, the only Remedy, that can be found for this *extravagance:* and that has been a constant Resolution, to reject all the amplifications, disgressions, and swellings of style: to return back to the primitive purity, and shortness, when men deliver'd so many *things,* almost in an equal number of *words.* They have exacted from all their members, a close, naked, natural way of speaking; positive expression; clear senses; a native easiness: bringing all things as near the Mathematical plainness, as they can: and preferring the language of Artisans, Countrymen, and Merchants before that, of Wits, or Scholars.

Croll conceives of the writer as a creature of nature and thus "natural," where Jones conceives of a writing as natural and thus as colorless as other objects of the Newtonian cosmos. Nature had to be conceived on a scale which could include a natural subject and a natural object. For one, style was the man, and, for the other, a completely styleless language was the equivalent of the objective world. Both speak of essential qualities of this new linguistic form; its allowance of equivocation was a necessary counterpart to the rigors of formal languages.

The prose which developed in the seventeenth century and which continued as the primary medium of knowledge cannot be simply defined. In a sense, its usefulness depends upon its lack of definition. It is the material from which both the formal language of mathematics and the formal language of poetry is refined. Of a passage from Pascal's essay on "Imagination," Croll says, "Nothing could better illustrate the 'order of nature ... ,'" which is an order of definite objects but uncertain relationships, relationships which are—to use a words that has been common in recent discussions—subject to "sliding or overturning of former categories." Prose elegantly allows these imprecisions which are excluded from both formal mathematical languages and poetry. Croll goes on to note that the sentence

> begins by naming the subject, the *plus grand philosophe,* without foreseeing the syntax by which it is to continue. Then it throws in the

elements of the situation, using any syntax that suggests itself at the moment, proceeding with perfect dramatic sequence, but wholly without logical sequence, until at last the sentence has lost touch with its stated subject. It is a violent, or rather nonchalant, anacoluthon.

This is a fair characterization of one boundary of the "world of prose" which, according to Hegel's analysis, appears in the nineteenth century. The other boundary is formed by a normative discipline—a rigorous grammar or empirical investigations, Cartesianism or Baconianism. In effect, the resources of abstraction which mathematical analysis supplied required an entire culture as its medium. It was the function of prose to communicate with the incommensurable elements which were necessarily involved in the synthesis—to translate, to absorb the slippage, play, and so forth. As Robert Adolph writes, "From the Restoration on, normal literary prose is, to use McLuhan's terms, a 'linear' product of the 'print culture.' The chief aim of such prose is useful public communication." It became an instructional medium, devoted, as Stanley Fish notes, to the universal belief of the Restoration in "the ability of the mind to be instructed in the truth." There were some specialized vocabularies, of course, which distinguished certain professional groups, but prose itself became transparent and is only now called into question.

2. The Two Languages: Space

For the modern scientist energy has no borders, it is a shapeless 'mass' of force; even his capacity to differentiate it to a degree never dreamed by the ancients has not led him to think of its shape or even its loci.... Perhaps algebra has queered our geometry. —Ezra Pound

Compendious Abbreviations

Jacques Derrida writes: "The 'formal essence' of the thing can only be determined in terms of presence. One cannot get around that response, except by challenging the very form of the question and beginning to think that the sign is that ill-named thing, the only one, that escapes the instituting question of philosophy: 'what is?' Of the classical tradition, Derrida's assessment is no doubt accurate. In modern philosophy, however, presence is always and only a sentimentality, albeit a necessary sentimentality, until it was, over a period of centuries, eroded away and replaced with its replica—a statistically fulgent zone in the media flux which can now be controlled by the machinations of popular culture rather than by metaphysics and theology in collusion with art.

Why Derrida's proposal of a grammatology has aroused such exaggerated interest at this late date is not clear. A deep reactionary spirit must need to believe that the ontological question is still open, that the sentimentality which has invested "self," "imagination," and "environment" with their powerful resonances can be sustained, if only by negation. But Descartes set a process into operation which moved us inexorably toward modernity. It is only the profoundest incoherence—incoherence invoked on behalf of an all-embracing totality—that breaks the sway of an old mentality and allows the movement toward the new.

From his earliest work—the suppressed essay *The World; or Essay on Light*—Descartes assumes a thoroughly literate cosmos in which voice, like sight *per se*, has no significant relation to the abstract objects which constitute the field of knowledge:

> You know that words, while having no resemblance to the things which they signify, do not fail to make them intelligible to us, and often, even without our paying attention to the sound of the words, or to their syllables; so that it may happen that after having listened to a discourse, the meaning of which we have completely understood, we are not able to say in which language it was spoken. But if words, which signify nothing except by human institution, are capable of making conceivable for us things to which they have no resemblance, why may not nature also have established a certain sign which should make us feel the sensation of light, although this sign should have nothing in itself resembling sensation? Has she not thus appointed laughter and tears to make us read joy and sadness in the human countenance....
>
> A man opens his mouth, moves his tongue, expels his breath; I see nothing in all these motions which is not quite different from the idea of the sound which they cause us to imagine. And most philosophers assure us that the sound is nothing but a certain trembling of the air which has just struck our ears; so that, if the sense of hearing brought to our thought the true image of its object, it would be necessary, in place of making us conceive of the sound, that it should make us conceive the motion of the portions of the air which is trembling at the time against our ears.

The voice bespeaks itself in language which is not heard, nature writes itself in signs which have no relation to the object, and the soul writes its emotions on the human countenance in signs which must be interpreted. Descartes' dualism divides meaning from voice as decisively as it does body from soul.

The central concern of Descartes' *Rules for the Regulation of the Mind* is the relationship between the common language and the conventions of mathematical formalism. His project relates directly to those of his contemporary linguistic projectors. He hopes to translate from the phonetic language to an algebraic language of geometric glyphs. In this connection, he recognizes the need for "certain compendious abbreviations" which supplement memory and aid "the continuous and uninterrupted action of a mind that has a clear vision of each step in the process." It is not an extreme exaggeration to say that the Cartesian method is fundamentally typographical: it is concerned with graphic arrangements which clearly reveal structure of evidence, so proofs are easily read and the relationships between ideas made clearly manifest. To think in terms of contemporary mnemonic technology, the rules for the regulation of the mind address the problem of programming, the creation of software for the printing press. Descartes' proposal borders on a mysticism of inscription: "... since [memory] is liable to fail us and in order to obviate the need of expending any part of our attention in refreshing it, while we are engaged with other thoughts, art has invented the device of writing." And, he advises: "nothing that does not require to be continuously borne in mind ought to be committed to memory, if we can set it down on paper."

What does the method require to be continuously borne in mind? Or, conversely, what cannot be committed to the formalism? These are practical questions, the kind of questions which must arise for a practicing mathematician, and questions of this kind, more significantly than questions of ontology, have motivated the history of the past three centuries and over. If the intuition and the expression of mathematical entities required the same kinds of objects, the questions would be simply answered, but they do not. The content of "naked understanding," in Descartes' terms, is "simple natures," entities as colorless and as empty as the natural integers which can be arranged in sequence by a simple enumerative process and then combined by rules to produce all possible, valid combinations. For Descartes, these entities may be spiritual, corporeal, or *both at once*. He appeals to them first as ideal objects which must "be taken altogether outside the bounds of the imagination, if they are to be true." And again as merely pragmatic considerations: "It matters little, however, though they [the simple natures] are not believed to be more real than those imaginary circles by means of which Astronomers describe their phenomena, provided that you employ them to aid you in discerning in each particular case what sort of knowledge is true and what false." There is an inherent contradiction in the idea of an imaginary ideal—that is, an imaginary object of which no image is possible. It is, however, precisely this imageless—which is also to say measureless—image which Descartes takes as the very building block of his system, and which becomes the mark of modern thought. It is variously the infinitesimal point of mathematical analysis, the dimensionless point of difference in linguistics.

Once marks are allowed to stand as images for simple natures, as Derrida is accurate to note, metaphysical purity necessitates that the entire argument be comprehended in a single view, as "present to itself." Descartes, however, explicitly makes no such claim: "... what I have to do is to run over them [i.e. the properly concatenated propositions in the proof] all repeatedly in my mind, until I pass so quickly from the first to the last that practically no step is left to the memory, and *I seem to view the whole all at the same time.*" He is careful to stay within the *practical* limits of the method: the instantaneous apprehension of the whole which absolute knowledge requires is temporalized. He writes: "... even though the understanding in the strict sense attends merely to what is signified by the name, the imagination nevertheless ought to fashion a correct image of the object, in order that the very understanding itself may be able to fix upon other features belonging to it that are not expressed by the name in question." The ability to relate ideal simple natures to images—literally, marks on the page—is the essential equivocation of modern knowledge. The imagination integrates the names, which for "naked understanding" are in the realm of abstract potentiality, in its world-constituting function. In the century after Descartes, this practical function was sentimentalized and given responsibility for articulating an aesthetic world—the domain of artists and engineers alike.

The evidence for the "unfailing complicity ... between idealization and speech," in Derrida's words, simply does not appear in Descartes or the work of the other essential thinkers—that is, the thinkers who contribute directly to the development of contemporary technoculture. If we emphasize those features of methodology which underwrite its efficacy rather than those which attempt vainly to maintain its consistency, the complicity between idealization and *writing* is clear, and, when voice *is* idealized, typically its structure is bent to the demands of phonetic writing and not the reverse. Certain primitivists, Rousseau and Lévi-Strauss, whom Derrida takes as typical, have invoked the classical tradition on behalf non-literate people, and in a twentieth-century attempt to salvage the Cartesian project, Edmund Husserl idealizes voice as Derrida elegantly demonstrates in *Speech and Phenomena*, but these are anomalous examples. To the extent that Husserl idealizes voice, he follows neither Descartes' lead, as we have seen, nor the lead of the other dominant strain of modern philosophy—that is, Locke's, with its insistence upon the primacy of writing. Locke's metaphor for the mind as a *tabula rasa* underscores the dominance of literacy in the modern *episteme*. We have had a science of writing and *only* a science of writing for the past three-and-a-half centuries.

The ideal self and the ideal object were not required by the Cartesian project; they were rather sentimentalities, concessions to the human need for certainty, not to the practical requirements of the methodolgy. The power of the formalism, confirmed by the successes of technology, is now secure, and the fact that neither self nor object were functional is exposed. We are all Cartesian modernists, who, by staging an idealism as the pretext for a language without etymologies or sensuous content, have undergone a new beginning. The proliferating formalism often seems overwhelming. It appears to have no outside and no concrete content. Increasingly we are its creatures. If we can expose clearly its implementation, to see how we gradually replaced our intuition of the world and its measured language with a logic and an aesthetic, however, we can begin to reclaim a concrete common world as the content of a community.

Intuition

The seventeenth-century epistemologists reordered the relationship between the arts and the sciences as they were understood in classical and medieval times. In the classical view, art emphasized the practicality of judgment; science emphasized theory for its own sake. Although the tradition of philosophic utopianism proposed to apply theory to politics (which was the highest art), the thought of applying theory to nature in a practical sense never occurred to the classical thinkers. The Cartesian position represents a change in desire, a reassessment of the relation between thought and the physical world. In the *Tractatus*, the last great document of the Cartesian tradition, Wittgenstein says: "anyone who understands me eventually recognizes ... [my propositions] as nonsensical, when he has used them—as steps—to climb up beyond them. (He must, so to speak, throw away the ladder after he has climbed up it.)" In the first four rules of the *Regulae*, we witness Descartes' somewhat inelegant scramble up his own ladder of non-sense. Commentators have tried to explain his incoherence, but we must recognize that radical modernism involves not incoherence as a mere intellectual error but as a consuming and powerful intellectual practice. Thus, the logical impossibility of beginning again is overcome.

The passages on intuition in the *Regulae* are some of the last tentative looks of Faustian man, self-creating man, whatever we should call the modern human, back to the firmly-rooted ontology of the classical world. Descartes substitutes "clear and distinct ideas" for "intuition" in his later work, and when the term returned to philosophic currency in the work of Kant, it had a different valence; it no longer related to objects as such but to the conditions which make the apprehension of objects possible—that is, the forms of sensibility, time, and space. Descartes' definition of "intuition"—in its particular unclarity—is revealing:

> By intuition I understand, not the fluctuating faith of the senses or the fallacious judgment of a badly compounding imagination, but the conception of a pure and attentive mind, so distinct, that absolutely no doubt remains about what we are understanding; or, what is the same thing, the undoubting conception of a pure and attentive mind, which is born of the light of reason alone, and is more certain even than deduction, because simpler, even though we have already noted that the latter cannot be badly performed by man.

We learn a good deal more about what intuition is not than what it is: it is not sense perception, it is not imagination, and it is not deduction. Although it is not reason as such, it is "born of the light of reason"—that is, apparently, to the mind which *shines* with reason, purely and attentively. His examples are of two sorts, intuitions of the self and intuitions of geometric properties: "Thus each individual can mentally have intuitions of the fact that he exists, and that he thinks; that the triangle is bounded by three lines only, the sphere by a single superficies, and so on." Thus, we have knowledge of at least one singular object, the knowing self, and of certain classes of objects which are tautologies—the necessary constituents of judgments which are both synthetic and *a priori*. The problem for epistemology was to join them. The argument which Kant would finally articulate is implicit in Descartes' development of the idea.

Rule 1 states: "The end of study should be to direct the mind towards the enunciation of sound and correct judgments on all matters that come before it." "Judgment" is a common term of traditional logic which has to do with building up true propositions from simple elements. It is the faculty which synthesizes complex truths from simple natures. The notion, however, that the *focus* of intellectual effort should be judgment rather than syllogistics was radically new. According to the methodologists, medieval logic had been lax in judgment, that is, in determining the truth of the premises. At once paraphrasing and revising Aristotle in his commentary, Descartes writes, "sciences . . . entirely consist in the cognitive exercise of the mind, [and] . . . the arts . . . depend upon an exercise and disposition of the body." The arts, with their physical involvements in the world, deal with *particular* subject matters. To his first examples, agricultural operations and harpplaying, he adds investigations into "human customs . . . the virtues of plants, the motions of stars, the transmutations of metals," and so forth. The sciences, on the other hand, "taken all together are identical with human wisdom, which always remains one and the same, however applied to different subjects, and suffers no more differentiation proceeding from them than the light of sun experiences from the variety of the things which it illumines." The science for which Descartes speaks represents renaissance megalomania at its most dramatic: it has no specific content; it is rather the rules by which any possible content can be constructed; and, above all, it empowers reason, "not for the purpose of resolving this or that difficulty of scholastic type, but in order that his understanding may light his will to its proper chance in all the contingencies of life." The moderns propose to deal with the contingencies of life from the perspective of that which is *not* contingent and to renovate the physical world on behalf of the mind. Knowledge of particular things and events derives from judgments of universals, so the actual world is secondary to the Cartesian ego. "Judgment" is no longer a matter of building up isolated propositions from simple apprehensions; it is rather a world-constituting act. The first proposition of modern philosophy is technology.

Western science is neither simply a method nor a body of knowledge but a complex institution which coordinates method and knowledge with psychological forms. I know of no scholarly account of the crucial role which art, and especially the timed arts—poetry and music—played in the development of the scientific project of the seventeenth and eighteenth centuries. They too became in a sense technological, implementations of a soul. Descartes set the stage for a powerful movement in art which created the requisite psychology of selfhood. In Milton and Donne, in El Greco, in Bach, and in the facades of Baroque churches, we can see signs of a new human creature which understands itself as infinite potentiality, as the generator of language which contains all knowledge and which expresses not the outcome of science but its resources. It was the artist who supplied the *imagination* which methodology required as a substitute for intuition. The new mode is clear, for example, in Bruno's attacks on Petrarch, and in the wide-spread appearance of metaphysical poetry, "Concettismo," "Marinismo," "Gongorismo." The new images were not based on similitude. The vehicles were intuitable, but the relationships between them were not. At *some* level of abstraction, everything is like everything else. The modern mind was more at home with paradoxes than with ontological mystery; it preferred recursive languages which require tireless but potentially entertaining interpretation to an hierarchical language which requires absorbed contemplation. Baroque art was the aesthetic leap into the beyond which corresponded directly to mathematical analysis.

The Equivocations

The difficulties which are presented by the self-reflective requirements of methodology appear in the commentary on Rule IV: "There is need of a method for finding out the truth." In itself, this rule does not seem powerful but it goes directly to the heart of the modernist objection to scholastic learning which made heavy concessions to contingency. The self-reflective nature of modern intuition is contrasted to mimetic or classical knowledge.

> So blind is the curiosity by which mortals are possessed, that they often conduct their minds along unexplored routes, having no reason to hope for success, but merely being willing to risk the experiment of finding whether the truth they seek lies there. As well might a man burning from an unintelligent desire to find treasure, continuously roam the street, seeking to find something that a passer by might have chanced to drop. This is the way most Chemists, many Geometricians, and Philosophers not a few prosecute their studies.

Descartes delineates a ludicrous character in this passage, but it betrays a complete lack of understanding for or sympathy with a primary form of intuition which is mimetic rather than methodological. To characterize this intuition as conceptual or perceptual, for example, is *already* to bias its possibility of dealing with matters in some wholeness. The mimetic sense of the world presupposes an object of imitation the nature of which is radically different from, perhaps even incommensurate with, the medium of imitation. Methodology institutes an inherently linguistic world as its field of operation.

Certain textual peculiarities in the *Regulae*, and especially in the commentary to rule IV, are revealing. It is possible to see Descartes struggling with the complexities of his incoherent strategy. Jean-Paul Weber was by no means the first to note certain inconsistencies and repetitions in the text, but he argues, largely on the basis of such evidence, that it is little more than a rough collection of passages from Descartes' youthful writings. He suggests that the work is a mere conflation of notes, written over a period of time, and embodying different conceptions of the method. The variant texts—or what appears to be such—cannot be as easily reconciled as Jean-Luc Marion suggests in his *L'ontologie grise de Descartes*, nor can they be explained as clear evidence of a crucial turn in Descartes' thought, as Pamela Kraus argues. The questions Weber raises are, to say the least, significant enough to require attention by anyone who would base an argument on the *Regulae*.

The first paragraph of the commentary to rule IV more or less recasts the first three rules, and the second asserts the rediscovery of an ancient mathematical method. Although we do not know exactly when the *Regulae* was written, it was probably after Descartes had discovered algebraic geometry, and those developments seem to be on his mind because he tells us that he is not thinking of an "ordinary mathematics" but rather an "instrument of knowledge ... [which is] the source of all others." The confusion over the nature of this science has been far greater than is justified by the textual difficulties. The problems are logical, not textual, and they remain a constant of the philosophic tradition from Descartes himself, who found a way in the *Discourse* to sweep it under the rug, to Noam Chomsky and the foremost current French proponent of geometric mechanism, René Thom.

Although most of the "repetitions" which Weber notes can be accounted for in terms of a perfectly normal thematic development which could have been worked out in revision, there are a few passages in which Descartes is clearly struggling with serious philosophic problems. Consider, for example, this pair of passages:

> 1. For the human mind has in it something that we may call divine, wherein are scattered the first germs of useful modes of thought. Consequently it often happens that however much neglected and choked by interfering studies they bear fruit of their own accord.
>
> 2. But I am convinced that certain primary germs of truth implanted by nature in human minds—though in our case the daily reading and hearing of innumerable diverse errors stifle them—had a very great vitality in that rude and unsophisticated age of the ancient world.

These are not sentences of an inept or immature writer; they are written with a sensitivity to tone, nuance, and rhetorical impact. Whether the germs of truth are divine or natural might seem a matter of significant indecision, but it makes no difference to the actual *structure* of the argument. In either case they are *innate* and, equally important, they were known to the ancients; that is, a trace of them exists in the traditional learning. The *problem* is accounting for error. The tradition is obviously responsible for the "diverse errors" of the schoolmen, but it must, at the same time, somehow give evidence of these originary germs of learning; that is, the old house of the new culture cannot be entirely disregarded.

The same problem emerges dramatically in his account of the relationship between ancient and modern mathematics. Again, in these passages, Descartes is trying out alternate assessments of the modern renovation:

> 1. At the present day also there flourishes a certain kind of arithmetic, called Algebra, which designs to effect, when dealing with numbers, what the ancients achieved in the matter of figures.
>
> 2. For it seems to be precisely that science known by the barbarous name Algebra, if only we could extricate it from that vast array of numbers and inexplicable figures by which it is overwhelmed, so that it might display the clearness and simplicity which, we imagine, ought to exist in genuine mathematics.

As in the previous passages, Descartes' stance is already in the schematic world picture which depends upon a new formal language for its expression. If there is only the mechanism of unreasoning habit on the one hand, and the invariant rules of the *mathesis universalis* on the other, how does one make the transition from the false mind to the true? If knowledge is innate, it is impossible to account for the pervasiveness of error. If, however, it represents genuine innovation, it is impossible, having only the false tradition, to account for the ability to recognize the new truth.

These inconsistencies represent the efforts of a major philosopher, entering the mature phase of his thought, struggling with an insoluble problem at the center of his system. The fact that the system has had pervasive influence on the subsequent intellectual history of the West should perhaps be enough to make us doubt whether logical consistency is as fundamental as we have believed. In terms of the increase of intellectual efficacy, in fact, the evidence might more fully support the usefulness of allowing thought to play back and forth between a willful dualism and a theoretical commitment to coherence. Any ideal language necessarily waffles in relationship to the common language it is intended to replace. The relationship between methodological thought and colloquial thought, between the new mathematics and the scholastic tradition, is necessary and necessarily obscure. Had Descartes prepared the manuscript for publication, he would have no doubt produced a text in which the problems would not have been so blatant, but the equivocation is essential to the program itself.

Rules V and VI propose a method for analyzing complex arguments into their constituent parts and ordering simple natures in useable form, while Rules VII through XI outline synthetic techniques for reconstituting the world in an ideal mathematical language. Rule VI, in particular, Descartes tells us, contains "the chief secret of method.... For it tells us that all facts can be arranged in certain series, not indeed in the sense of being referred to some ontological genus such as the categories employed by philosophers in their classification, but in so far as certain truths can be known from others." By carefully keeping track of absolute and relative terms—the structure of the hierarchy—it is possible to enumerate statements in order from the simplest and clearest to "whatever is said to be dependent, or an effect, composite, particular, many, unequal, unlike, oblique, etc." By exploiting the mnemonic technology of writing and particularly writing in the convenient and easily manipulated form of print, data can be managed in sure and powerful ways.

If the analytic function is the great secret of the Cartesian method, Descartes gives a much fuller and less confusing account of the synthetic process. It is not necessary, for our purposes, to go into the procedures; more sophisticated versions of them are known to undergraduate mathematicians. The kind of *attention* which is being proposed beginning in Rule VII, however, requires a radically new psychology. If thought can be regulated in its own operations and thereby released from the contingency of world, it can explore all possibilities. *Classical thought attended to the relation of form to thing; the modern to the relation of form to form.* Descartes writes: "I am now able by attentive reflection to understand what is the form involved by all questions that can be propounded about the proportions or relations of things, and the order in which they should be investigated; and this discovery embraces the sum of the entire science of Pure Mathematics." The confidence that the mind can extend its knowledge of actual structure to the structure of actuality is essential to the modern will to power.

In Rule XII, which summarizes and concludes the first section of the *Regulae*, Descartes reintroduces the use of imagination, sense, and memory, but now their subordination to intuition and deduction—or the understanding, as he now calls them—is assured:

> Finally we ought to employ all the aids of understanding, imagination, sense and memory, first for the purpose of having a distinct intuition of simple propositions; partly also in order to compare the propositions to be proved with those we know already, so that we may be able to recognize their truth; partly also in order to discover the truths, which should be compared with each other so that nothing may be left lacking on which human industry may exercise itself.

The mechanical quality of imagination, sense, and memory is emphasized, and the image of the human which Descartes presents is a being, consisting of two texts, or as existing at the intersection of two texts, one which belongs wholly to the physical, mechanical world, and one which is divinely or naturally inscribed in the mind in the form of intuited simple natures and is supplied with its own techniques for combining to encompass descriptions of all of the world's possibilities.

In a metaphor which he carries through an extended passage, Descartes compares the body with its secondary intellectual faculties to a writing pen. In early modern grammatology, this passage must stand with Locke's discussion of the *tabula rasa*. Judgment is always judgment of a specific writing. He initially introduces the figure to explain how sense impressions are transmitted from the external world to some part of the body without the passage of "any real entity from one to the other":

> It is in exactly the same manner that now when I write I recognize at the very moment when the separate characters are being

written down on the paper, that part is simultaneously shared by the whole pen. All these diverse motions are traced by the upper end of the pen likewise in the air, although I do not conceive of anything real passing from the one extremity to the other. Now who imagines that the connection between the different parts of the human body is slighter than that between the ends of a pen, and what simpler way of expressing this could be found?

Conversely, the fancy manifests itself by the opposite motion of the pen.

This example also shows how the fancy can be the cause of any motions in the nerves, motions of which, however, it does not have the images stamped upon it, possessing only certain other images from which these latter follow. Just so the whole pen does not move exactly in the way in which its lower end does; nay the greater part seems to have a motion that is quite different from and contrary to that of the other.

It is unfortunate that the discussion of Descartes' mechanism has focused so heavily on the question of animals. Although his readers have been scandalized by the thought that horses and dogs may be automata, Descartes thinks of the larger part of *human* behavior as mechanistic as well. He argues that language is necessary to rational behavior and that the lack of language in animals is sufficient to prove that their actions are mechanical, but he does not say that language *assures* rational behavior. The activities of mind are rare and occur only when the germs of knowledge are awakened and utilized. They may, and presumably did, lay dormant for eons, and even in times when the method is known, true understanding is uncommon. The casual capacity for language—that one has a store of words, grammatical patterns, certain habitual uses, etc.—may be merely habitual and automatic. The other requirement of rational behavior is *method*. The effect of the Cartesian compromise is the identification of meaning with pattern or, one could say, information. Intelligence has been directed almost exclusively at pattern recognition and the kinds of predictions which it makes possible. The meaning of 'meaning' has been reduced to a sense of probability that a given linguistic event will recur, as the grain of a texture.

Descartes' rhetorical strategy in the *Discourse* is to stage the epistemological question in terms of autobiography, thus shifting the question from matters of theoretical consistency to matters of education: "... I shall endeavor in this discourse to describe the paths I have followed, and to delineate my life, in order that each one may be able to judge of them for himself...," and so forth. Consequently, he is never required to translate whole-cloth from the common language to the ideal. The process, rather, is the initiation into error which is gradually recognized and corrected. The circularity of the argument is less obvious, because we allow that people learn from their mistakes, while logical systems do not. It is the prototype of other self-improvement schemes. Since Descartes, we have increasingly placed our trust in our ability to construct the world (and self) we require, first as mathematical replicas in the seventeenth and eighteenth centuries, then as technological replicas in the nineteenth and twentieth centuries.

Linguistics

Prose posits its systematicity and seeks to disclose it. Although Noam Chomsky's historical research on the Cartesian tradition has been reviewed harshly, he reveals an essential fact of the understanding of language in the modern West. Perhaps we can think of his writings in *Cartesian Linguistics* and in the first and third chapters of *Language and Mind* as more mythological than scholarly; he reveals our linguistic unconscious. The thought of Descartes' "primary germs of truth implanted by nature in human minds" is as essential to the culture that built the World Trade Center, the spaceship Columbia, and Disneyland as Athena is to the culture that built the Parthenon, or as the Platonic myth of Er, the Pamphylian, is to the declining Greek empire. Chomsky describes language as a mode of free cognitive creation, and the Cartesian precedent for his argument is strong. "Descartes ... described human reason as a universal instrument which can serve for all contingencies and which therefore provides for unbounded diversity of free thought and action."

Chomsky takes little note, however, of the distinction between natural language and *mathesis universalis* which, for Descartes, constituted a significant boundary to the freedom. Indeed the passage which Chomsky cites implies that *at least some* language is bounded, unfree, and reactive. Descartes writes: "... mind is of such a nature that from the motion of the body alone the various sensations can be excited in it," and as we have seen, language without careful methodological discipline is a manifestation of physical mechanism. In the commentary to this principle, Descartes goes on to say:

> We observe that words, whether uttered by the voice or merely written, excite in our minds all sorts of thoughts and emotions... we

can trace letters which bring to the minds of our readers thoughts of battles, tempests or furies, and the emotions of indignation and sadness; while if the pen be moved in another way, ... thoughts may be given of quite a different kind, viz. those of quietude, peace, pleasantness, and the quite opposite passions of love and joy.

Language in this sense, which is of course precisely the creative mode of poetry as it is generally understood in romantic theory, is according to Descartes mechanical, and no doubt the emphasis on the pen, the mechanical instrument of language, functions rhetorically to underscore its mechanical nature. Although Descartes never doubts the rich creativity of language, the problem, as he understands it, is to find a unity of language which is not constantly subverting itself with something new or compromising itself with mechanism. The Cartesian method is directed toward rational control of this unruly linguistic machine which generates its own illusory world.

Chomsky reads Romantic linguistics back into Descartes. In retrospect, however, Chomsky's assessment of the general spirit and direction of Cartesianism is sound. Descartes' classicism was a requirement of his instigating project only because he had to stage an entire world for his ego. In time, the culture would provide the necessary space, and the functions of intuition could be passed first to imaginative literature and then to the popular media.

3. The Two Languages: Time

In English the poetics became meubles—furniture— thereafter (after 1630

& Descartes was the value

—Charles Olson

The Dream of the Cyborg

Quickly to focus the issues, Descartes substituted mathematical analysis for the external world. Newton, recognizing the mistake, tried to reverse the field, and he turned Descartes inside out—as later Marx would turn Hegel upside down—but it was too late. Substituting his own empiricism for Descartes' rationalism, Newton changed the perspective but not the metaphoric structure: he transferred the divinely-underwritten logical structure of ego to a prior and absolute space which he identified with divinity. That is, the founders of modern physics, despite their apparent oppositions, both conceived of the difference between the human and the divine as a matter of perspective. The chief intellectual project thereafter was the creation of human replicas to inhabit these perspectives which were oddly more rational than humans themselves. Descartes even conceived of constructing automata which might appear fully human:

> For we can well imagine a machine so made that it utters words and even, in a few cases, words pertaining specifically to some actions that affect it physically. For instance, if you touch one in a certain place, it might ask what you want to say, while if touched in another, it might cry out you're hurting it, and so on.

Descartes' model, which treats the nervous system in terms of coding and information transfer, is a direct forerunner of contemporary models in cognitive science and artificial intelligence. Descartes could not, however, account mechanistically for functions which are considered "mental," that is, especially *purposive* functions, and he goes on to say, "... no such machine could ever arrange its words in various different ways so as to respond to the sense of whatever is said in its presence—as even the dullest people do." His automata might have behaved as if they were doubting, thinking, and knowing, but they could not doubt, think, and know.

Until this century, it was not clear how to reconcile purposive and mechanistic behaviors; they have different temporal structures. Cartesian science related the adequacy of causal explanation to the atemporality of geometry, thereby displacing scholastic teleology from the ontological to the cultural domain and rendering questions of purpose merely pragmatic. The cultural institutions at large became responsible for managing change or "progress" in a concrete realm which was implicitly technological; that is, teleology became a cultural and historical project. The human agent as a purposive motor became the supreme cultural product. The function of education and the arts was to create progress and account for a history which was inexplicable in terms of our most powerful and useful understanding of physical nature.

Poetry and Mathematics

The poetry of the seventeenth century is self-referential, as its mathematical logic is tautological. It avoids the ultimate nihilism which is implicit in its project by proposing ever-larger linguistic fields. Language is larger than any "I" which expresses itself by words, and the pursuit of self-expression through language leads systematically, by the logic of language as such, to knowledge of the absolute *and* self-alienation. If one sees language as a whole from an eternal perspective, as Descartes saw geometry, the entire fantasy collapses: I am I, everything is only what it is, integral and meaningless. Thus, the tradition of poetry which arose among Descartes' contemporaries insists paradoxically upon finitude as a way of energizing the ideal. It was a self-exploitation from the beginning and could only, as Hegel was accurate to note, sublate itself.

The most significant "essay" not on, but into, this matter is Robert Duncan's "A Seventeenth Century Suite in Homage to the Metaphysical Genius in English Poetry (1590–1690): Being Imitations, Derivations & Variations upon Certain Conceits and Findings Made among Strong Lines," in *Ground Work: Before the War.* Proposing an homage, Duncan enters fully into the Metaphysical spirit and reveals both its interior richness and its lack of grounding in anything but itself. He inhabits a limited zone of poetic imagery, which is not most immediately attractive to him and sets it against the demands of his own poetics, discovering the uses and the beauties of seventeenth-century poetic strategies and, at the same time, locating and exceeding their limits. "A Seventeenth Century Suite" is not only a remarkable set of poems, it is also a critical act of the highest order.

Many of the poets of Duncan's generation had written in imitation of seventeenth-century poetry, intending only to continue the cultural imperative, which had worn itself out in the nineteenth century, to imagine an ideal self. Duncan, however, undertakes the project as one of the most articulate polemicists against the old New Criticism which had sponsored this kind of poetry. In his earlier essays, he had objected to the conventional form of the neo-metaphysicals, but it was not the pentameters and the quatrains *per se* that drew his ire; it was the use of conventional poetic techniques to enforce normative morality and normative feeling:

> Form to the mind obsessed by convention, is significant in so far as it shows control.... Wherever the feeling of control is lost, the feeling of form is lost. The reality of the world and men's habit must be constricted to a realm—a court or a salon or a rationale—excluding whatever is feared. It is a magic that still survives in Christian Science and the New Criticism, a magic that removes the reasonable thing from its swarming background of unreason—

unmentionable areas where all the facts that reason cannot regulate are excluded and appear as error, savage tribes, superstitions and anarchical mobs, passions, madnesses, enthusiasms and bad manners.

In "The Seventeenth Century Suite," recognizing that language itself is conventional, Duncan addresses convention without obsession. He knows that to honor a particular convention, even by denying its validity, is to privilege one convention against another and to create idealisms in the otherwise open field of language. He demonstrates in his seventeenth-century workings that his pluralism is large enough to include the metaphysical impulse which he had rejected as the "official" style of New Criticism. The field in which the mature Duncan's work conducts itself is large enough to contain conventional selves, such as speak in seventeenth-century poems *and* the vast spaces exterior to them. The space can be explored, as it were, from both sides. Duncan exploits the aesthetic space and then steps outside it so we see zones of particular kinds of intensity arise and dissipate into larger and more serviceable fields.

Pedants will object that Duncan uses the term "metaphysical" inaccurately, but they will only show that he does not make their mistake of confusing styles with poetics. From Duncan's point of view, all of the significant poetic styles of the seventeenth century are implicated in the substitution, which is the work of the metaphysical genius, of a conventional imagination for intuition: poets turned their attentions to the relationships of image to image, of form to form, of language to language, their stylizations paralleling those of the *mathesis universalis*. Both the most extravagant devotees of metaphysical conceits and the neoclassicists sought ways to regulate the common language by the idealizations of verse.

Courtliness and Recursion

In the prelude to the suite, "Love's a great courtesy to be declared," Duncan positions himself in relation to the conventions not of metaphysical poetry, as one would expect, but of courtly love. That is, he addresses the seventeenth century from the medieval perspective, a perspective to which he returns in the Dante variations later in *Ground Work*. Medievalism is a no-less-conventional zone of poetic intensity, but it is larger than the metaphysical because it opens immediately outward: the ego of the courtly lover is unstable and therefore declares a space of unknown properties; its boundaries are not significantly dependent upon aesthetic structures. Duncan hearkens back with Ezra Pound, who speaks in his essay, "Cavalcanti: Medievalism," not of "pagan worship of strength, nor the Greek perception of visual non-animate plastic, or plastic in which the being animate was not the main and principal quality, but this 'harmony in the sentience' or harmony *of* the sentient, where the thought has its demarcation, the substance its *virtu,* where stupid men have not reduced all 'energy' to unbounded undistinguished abstraction." In the medieval perspective which Duncan initially adopts, the world of surfaces, of plane geometries, of linguistic conventions, are fields of contingency from which vision potentially arises. In the prefatory poem, he writes:

> I'd
> dissolve my soul in sleeping surfaces
>   where transient phantasies may come and go
> that somewhere in that multiplicity of
>   chance encounters
> I might come again to you and find
>   Love's court
> set up once more to rule my mind.

The issue is not a Cartesian regulation. Speech is its own catastrophe, dissolving souls in sleeping surfaces of conventions of the seventeenth century or of the poet's own time. To speak at all, Duncan knows, is to speak of wonders.

In *The Spirit of Romance,* which is an important context of Duncan's poetics, Pound asks, "Did this 'chivalric love,' this exotic, take on mediumistic properties? Stimulated by the color or quality of emotion, did that 'color' take on forms interpretive of the divine order? Did it lead to an 'exteriorization of the sensibility,' an interpretation of the cosmos by feeling?" Questions to which Duncan has consistently given an affirmative and completely secular answer. As early as 1947, Duncan had composed a series of poems, published as *Medieval Scenes,* "to exhibit [as he notes in 1978] mediumistic powers as well as to reach the voice of an oracle beyond that I performed":

> For me, it is the immediate appearance of an "other world"—not here, the Astral World, of departed spirits and divine beings, that my parents believed in, but the world Spicer and I called Poetry—that is most significant. I came upon the mode in which the eternal ones of the poem might speak to me. *Medieval Scenes* appeared in the medium of my writing at a *table parlant* where I consulted with fates that still stand over my work today.

In "A Seventeenth Century Suite," the immediate appearance of an "other world" is not the world of capital-P Poetry but the conventions of a kind of poetry which Duncan had once rejected as a model. It is an intentional transgression of his own boundaries, and no doubt a recognition that it is futile to assert one set of conventions against another. We are interested in the borders between different conventions where that which belongs to neither manifests itself.

The metaphysical genius is, above all, literate and literal, which is to say, it demands *everything* in writing. It is nowhere touched by contingency; everything is somewhere recorded. For the mark here there is a mark there: language is inside language, chasing its own tail. The image of lovers' bodies can be mapped point by point onto the image of a geometer's compass, and so forth. Language for the metaphysical poet calls forth more language and insists upon an immediacy of relationship even in the oblique occasions. Duncan opens the suite in the crisis of this recursivity, taking up Sir Walter Ralegh's theme from "What Is Our Life? a play of passions." Life imitates plays, a formula which must immediately turn upon itself, in that inside the play, plays imitate life which imitates plays, etc. It is a small and accurate parable of the infinite regress which has been the definitive modern event. The psychology is also characteristically modern:

> *Thus march we playing to our latest rest,*
> *Onely we dye in earnest, that's no jest.*

The fundamental change which had come about was clear to such men as Francis Bacon: modernity was "half in love with easeful death". In his essay, "On Death," Bacon writes: "You shall read in some of the friars' books of mortification that a man should think with himself what the pain is if he have but his finger's end pressed or tortured, and thereby imagine what the pains of death are, when the whole body is corrupted and dissolved; when many times death passeth with less pain than the torture of a limb; for the most vital parts are not the quickest of sense." The psychology which Freud outlines in *Beyond the Pleasure Principle* took hold in the seventeenth century. The homeostatic controls on history, created by an *intuited* world, had been removed, and the West had entered its period of wild growth, of proliferation not by mimesis but by methodology. The genetic and the personal have been utterly confused in mass society. To Freud the Cartesian combination of vital physicality and deathly mnemonics will seem to express itself as the erotic attraction of death. Duncan reads Ralegh as saying, "In death alone we are sincere," and implicitly we are cast between a groundless but attractive eros and the sincerity of death.

In the second variation, in which he retains Ralegh's theme but rejects his imagistic constraints, Duncan insists upon the knowledge which is missing in Ralegh's poem, of "deep uproilings/ of earth beneath your feet," roilings which, denied, are the powers of Kali, the Hindu goddess of destruction, whose emergence is one of the terms of the persistent sense of doom which haunts Duncan's work:

> Old dreams
> denied . the voided images go down
> into the preparation for catastrophe.

What Is

Perhaps one reason Duncan has not been more usefully read is his refusal to allow us the comfort of the literary. His task is to undermine the ego-security which the metaphysical poem proposes to create and to recover the knowledge which is repressed in them. In the variations on Robert Southwell's "The Burning Babe," he distinguishes between "a babe of fire" and "a baby on fire." The babe of fire belongs properly to the poem, as an image of the perfected self. There is no denial of these mysteries, which are the mysteries of self-consciousness and the persistent study of the modern West. Our obsessive questions have had to do with the nature of this recursive function: in Southwell's poem we read of "A pretty Babe ... such floods of tears did shed,/ As though his floods should quench his flames, which with his tears were bred." In "Imagination's alchemy," we await the return of the final deferred term which brings us to equilibrium. This is the conventional conclusion of modern mysticism, and, without denying the integrity of the vision, Duncan gives the passage an appropriately conventional mystical ending, drawing his terms from crucial passages in his own poetry:

> The burning Babe, the Rose,
> the wedding of the Moon and Sun,
> wherever in the World I read
> such Mysteries come to haunt the Mind,
> the Language of What Is and I
>
> are one.

Thus, Duncan recalls the language of an earlier poem which crucially defines the space of his work:

> Often I am permitted to return to a meadow
> as if it were a given property of the mind
> that certain bounds hold against chaos,
>
> that is a place of first permissions,
> everlasting omen of what is.

The conclusion, which is drawn in the Southwell variation, that "the Language of What Is and I/ are one" is one of the possible loci in the meadow: that is, it is a usurpation of the meadow on behalf of art. Of Southwell's babe of fire, Duncan writes:

> He's Art's epiphany of Art new born
> a Christ of Poetry, the burning spirit's show;
> he leaves no shadow, where he dances in the air,
>   of misery below.

However, Duncan can credit this "pretty babe" for only a short time:

> Another Christ, if he be, as we are,
> Man, cries out in utter misery.

Southwell's Christ exists only in the recursive symbolism of the poem. The mere prose and newspaper photographs which Duncan brings as evidence to the second Christ are irrelevant to the first. Here Duncan's testimony to multiple realities is at once cogent and somehow mind boggling: to think the world which is manifest here is to have something like the mental equivalence of the bends. In a lecture, written only two years or so before the first poems in "A Seventeenth Century Suite," Duncan had said:

> An ideal study of poetry would be concerned with all the kinds of poetries, the ideas men have had of what poetry is, with identifying the species of poetry, varieties of the poem in evolution.... We are concerned with World Order, and I would propose too that there is a nature or order of all poetries. But at best I experience only the different and differing orders of poetry that involve often incompatible ideas of what *world* and *order* are. Not only do we have different languages, we have different worlds and different orders; and within our American "world" and the particular language that the art of poetry creates there are communities of all kinds; each idea of poetry in so far as it is vitally concerned is charged with the conviction that it

Symbolic Nature

has a mission to change, to recreate, the heart of poetry itself. Each of us must be at strife with our own conviction on behalf of the multiplicity of convictions at work in poetry in order to give ourselves over to the art, to come to the idea of what the world of worlds or order of orders might be.

This passages seems to comment directly on the project of "A Seventeenth Century Suite." The true poets contend not only with opposing orders of poetry but with the order inside their own poetry as well. They are not bound by consistency, ideality, convention, or norms, *but by a world* which includes other poets, the vast array of differences and otherness.

In the next two poems, on George Herbert's "Jordan I" and "Jordan II," Duncan comes as near writing straightforward variations as he does anywhere in the suite. Herbert's professions of unruffled faith and love of simplicity bring forth comparably quiet, albeit completely secular, poems from Duncan—at least if they are read as isolated, organic entities rather than passages in the field of the suite and of Duncan's work. His models give no hint of self-division or internal strife. In "Jordan I," Herbert manages to oppose the whole tradition of fleshy, mythological and amorous poetry by saying, plainly, the words, "My God, My King." No poet had ever expected a simple naming to carry so much weight, and, when Duncan tries it, he sounds, in his first attempt, something like Polonius, and in his second like a skillful Iowa Writer's Workshop metaphysical. The intent, however, is not to appeal to taste but to bring forth certain evidence. In both cases, Duncan is able to suggest the recursive structures of the metaphysical imagination, and in both poems, he concludes with terms defined in terms of themselves:

> (I) Whatever I *believe*, my Art's to be true
> to what in truth is my Nature.

> (II) This water is but water. This is
>   no other water than it is, nor more nor less
>   that's meant to bless,
>   and works no magic
>   but our bliss.

The tautologies of mathematical logic are here mimicked in the poetic language which defines its terms recursively.

"Passages 36," the next poem in the sequence, belongs both to the suite and to the poetic domain of "Passages" which first appeared in

Bending the Bow. Beginning with a line from a dream, it is a direct consequence of the repressive requirements of the suite and a confrontation with the catastrophe which is prophesied in the second variation on Jonson's "What Is Life?" Grief is the mode of such knowledge:

> Is it to suit the myth yet to come—
> the ritual mutilation, the despoiling of nature, of earth,
>   of animal species, and mankind among them,
> with hatred and, no longer having a feeling of what is done,
> without hatred, day after day,
> the burning, the laying waste?
>
> Eat, eat this bread and be thankful
> it does not yet run with blood.

Like the Herbert poem, however, it ends with a powerful declaration of self-acceptance:

> I do not as the years go by grow tolerant
> of what I cannot share and what
> refuses me. There's that in me as fiercely beyond
> the remorse that eats me in its drive
> as evolution is in
> working out the courses of what will last.
>
> In Truth 'tis done. At last. I'll not
>
> repair.

This statement is as confident and assured as Herbert's; the difference is that it is without *particular* content. The "I" who speaks here can be defined only provisionally or negatively as an assertion of intolerance. Although the immediate context of the grief is "the end of an old friendship," it is grief for a world which has grown tolerant of its own refusal of itself, a world which closes itself off from the knowledge which can make its being tolerable. It picks up both of these concerns from an earlier poem in the volume, "Santa Cruz Propositions." There we read of the "Ur-Father, Hairy Bull of the Waters" who "was our Language come in to the Mothertongue." This change is associated in the poem with Socratic philosophy and especially the closing of language which was signaled by the relationship of Socrates and his abstract muse, Diotima. In the seventeenth-century models, however, the closure of language is complete, and, when they propose to manifest the truth, as in Ben Jonson's "Hymnaei," it reveals the poverty of its imagination:

Symbolic Nature

"Her orient hayre," I read:
"By which beleeving mortalls hold her fast,
"And in those golden chordes are carried even,
"Till with her breath she blowes them up to heaven."

Now what am I to do with that? tho I read there is a glow
where men's souls are quickened in her hair and
rise upon her breath toward heaven so, the poet's conceit
turns me back from the myth I know therein.

The figure of imagination which proposes to manifest Truth fails in its spectacle to make Truth any more than an advertisement, an obscene parody of the witness that it might bear to the love which the poem celebrates. The poem, however, does not make a simple point of taste. Jonson is given the most generous opportunity to bear his witness. The problem is not merely a matter of literary style or of failed poetry. Jonson and his contemporaries propose to situate us in relation to Truth, but, as Duncan writes:

> I do not know where I am with her,
>
> and myriad reflections upon her face
> lead from old deeps into new deeps of Night.

Since the seventeenth century, we have articulated a culture inside a language which, testifying on its own behalf, imagines the figures of its own truth. Jonson's lady Truth has, of course, had more attractive manifestations. Indeed she has been set forth in irresistable perfection, say, by Bach and Beethoven, where the world-defining strife is often neutralized in the logic of music. In his masterful study, *Beethoven and the Voice of God*, Wilfrid Mellers writes:

> In the tenth of his *Theosophical Questions* of 1624 Boehme asks, "What was it the devil desired, with a view to which he turned aside from God's love?" and replies: "He desired to be an Artist. He saw the creation and understood the foundation of it, whereupon he wished to be a God, and rule with the central fire-power in all things." It is in this sense that Beethoven, like Lucifer, became a *discoverer* of truth; and just as Bach's revelation of God turns out to be also a discovery of the Self, so Beethoven's discovery of the Self turns out to be also a revelation of God.

The Self-God/God-Self, the being which must include all, because it includes the language in which it is defined, is the figure of modern egotism. Some of Descartes' earliest critics noted that his argument was circular, that he took self-existence as the grounds for his argument for the existence of god and the existence of god as the grounds for his argument for self-existence. The mathematical Self-God and the poetic Self-God confirm one another. Perhaps it would be better to speak of the 'musical' Self-God as all of the timed arts are implicated. These are the great idealizations of space and time, geometry and music, which, contradicting and confirming one another in a careful dance, have required the participation of the entire culture. To support it has required untold exploitation of human and natural resources.

In the final poem of "The Seventeenth-Century Suite," Duncan takes up the theme of John Norris' "Hymne to Darkness." The turn to darkness is the only recourse for the poets who locate themselves outside of the destructive illumination, beauty, and quest for abstract power to which the culture devoted itself in the Seventeenth Century:

> yet striking ever true to what is
> most dark to me in me from that first
> darkend scale of all light Harmony
> asking, answering, note upon note of silent
>   command of tunings sound
> beyond sound.

This is the darkness from which the Self-God/God-Self in its divinity has withdrawn, which is explicitly identified in Norris' poem with the muse. It is also the darkness in which there is an Other and where love is therefore possible. In the last poem in the suite, the poet is startled in his meditation on the Lord's Prayer by a kiss from the darkness and the voice of his lover saying Good Night. The final line serves both as a conventional ending to the poet's enthrallment to a manner which is not completely congenial and as a reintroduction to his central theme. He notes completely without irony, "Love sets me free." The courtly pose of the opening poem is replaced with a domestic love relationship.

Real Time

Much as the images of the seventeenth-century poems have reference to other images and, ultimately, to language itself as the great Image of the World which is the content of Ralegh's play, the rhythms have reference to an *implicit* rhythmic structure in the language rather than the actual rhythms of the dance or the voice of the performer. In the standard formulation, typified by Wellek and Warren, "English verse is largely determined by the counterpoint between the imposed phrasing, the rhythmical impulse, and the actual speech rhythm conditioned by phrasal divisions." Despite a small skirmish in the 1950's between the proponents of conventional metrics and the linguists, who were interested in the description of actual poetic performances, this has been a more or less official doctrine of English metrics. The combatants, realizing that they were emphasizing different aspects of language, arrived at almost universal agreement that metrical forms involve the interplay of a conventional expectation and an actual expression. Northrop Frye, for example, states the compromise clearly:

> A four-stress line seems to be inherent in the structure of English language. It is the prevailing rhythm of the earlier poetry, though it changes its scheme from alliteration to rhyme in Middle English; it is the common rhythm of popular poetry in all periods, of ballads and of most nursery rhymes. In the ballad, the eight-six-eight-six quatrain is a continuous four-beat line, with a "rest" at the end of every other line. This principle of the rest, or a beat coming at a point of actual silence, was already established in Old English. The iambic pentameter provides a field of syncopation in which stress and meter can to some extent neutralize one another.

That is, the rhythmic *impulse* is understood to be imposed, as if it were foreign to actual speech as such. The confusion of rhythm and mechanical repetition in the culture at large is nowhere more apparent than in this standard doctrine. Rhythmic impulse in this sense can only be understood in terms of the reciprocating cylinder of the steam engine or some other mechanical device. The disappearance of voice, except as a soundless idealization, is complete.

In the present intellectual climate, it is difficult to raise issues of poetic technique in relation to knowledge. To suggest that rhythm is equally with image the content of knowledge meets with blank stares of incomprehension. In both poetry and music, rhythm is considered to be decoration. Duncan's revision of metaphysical verse, however, involves the restoration of impulse to voice—literal, physical voice, not the idealization of the metaphysical tradition—and the importance of his example is epistemological.

The essential epistemological mystery has traditionally involved the question of identity: how are things—the psyche, the tree, the other person, all of which constantly change—known as themselves? Now it is known that the question of identity has only a statistical, not an absolute, answer. The complement to statistical analysis requires attention not to identity and its absolutism but to invariant rhythms which may have long periodicities—involving perhaps entire lifetimes or even longer if we but knew how to record and study the patterns. That is, we must learn again with Heraclitus that only change is unchanging (the subtle dynamism of this formulation should not be confused with the deathly paradoxes which it resembles). The poets since Whitman and Melville have recovered at least primitive techniques of this knowledge.

One notices, reading Duncan's variations, that the rhythms are not only more varied and demanding of attention than the rhythms of his seventeenth-century models but they are of a different kind. All of the seventeenth-century poems are iambic pentameter, with the exception of Southwell's, and it is in fourteeners, a line which was created by running the eights and sixes of the ballad stanza together and regularizing the accentual verse with iambics. The strategy which creates the rhythmic effect of the fourteener is much the same as the strategy of pentameter: both recognize the inherent tendency in English to the four-beat phrase and create syncopations by imposing the expectation of one beat more or one beat less per line. By convention, syncopation is the mark of "serious" poetry. We immediately hear this verse, for example, as playful and childish:

> Hinx, minks, the old witch winks,
> the fat begins to fry,
> There's no one home but jumping Joan,
> father and mother and I.

As we shall see, Duncan's verse is *technically* nearer the nursery rime than the iambic-pentameter seventeenth-century poems: it does not invoke a generalized formal expectation which is independent of the particular performance of the poem. Our present concern, however, is with the relationship of seventeenth-century verse forms and the new intellectual methodology. In his important essay, "Problematizing the Pentameter," Anthony Easthope notes that iambic pentameter is implicitly ideological:

> The subject position [or the relation of subject to language in the terms I have been using] offered in pentameter, its effacement of poetic enunciation, its linear coherence, its repressive rigidity of closure can be referred to a traditional concept in literary criticism, that of tone. Pentameter aims to preclude shouting and improper

excitement; it enhances the poise of a moderate yet uplifted tone of voice, a single voice self-possessed, self-controlled, impersonally self-expressive, a tone which has retained its dominance in British culture since the Renaissance.

Easthope describes the voice of the ideal Cartesian self in its particularly sophisticated British form which, despite the efforts of our poets, has also been the mark of power in America. To be educated in poetry has been shamelessly to emulate this ideal of bourgeois speech which is at once powerful, self-possessed and, in terms of the other language of power, the formalism of science, more or less insignificant, except as the registrant of ultimate questions of ever-diminishing relevance. Significantly, in relation to Duncan's insistence that the true work of poetry is to embody and redeem all that it opposes, Easthope notes that "Brecht . . . attacks the imposed uniformity of iambic meter and protests against the ensuing 'smoothness and harmony of conventional poetry' which inhibits the showing of 'human dealings as contradictory.'" Brecht, to be sure, looks to some ultimate sublation of the contradictions in revolution, while Duncan understands them as a permanent fact of language and its appropriation of human agents to speak its forms.

Time is the dimension not only of cause but also of purpose: a purpose at time A grows into a plan at time B and issues as a success or failure at time C. In some views of time cause and purpose coincide. In medieval teleology, for example, purposive behavior was the instrument of cosmic purpose or natural causality: the purpose of an acorn was to become an oak tree and the purpose of a person on a journey to Paris was to get to Paris. At some point, implicitly, purposiveness would be used up, everything would have attained its nature, and time would end. This view of time was mythologized in the Christian doctrine of the Final Judgement: Christ would return to determine who had and who had not achieved their natures. Those who had proposed to achieve *themselves* rather than the cosmic design would be judged cosmic obstructions and, thus, would deserve destruction.

In the medieval view of time, cosmic time and the real time of poetic performance coincide. Augustine, for example, notes that time, while intuitively clear and obvious, cannot be well explained in language: "What, then, is time? If no one ask of me, I know; if I wish to explain to him who asks, I know not." Indeed as soon as we speak of time we are flirting with paradox: we speak of a "flow of time" but cannot account for loci other than the present in which time might be found; and we speak of the lengths of time but are baffled by the actual extension which we propose to measure. The conventions by which time is designated as present, as in "the present century" or "the present day" or "the present hour," involve all of the paradoxes of infinite divisibility. Some version of Zeno's problem inevitably lurks about considerations of time.

It is impossible to read Augustine's masterful analysis of the moment of present time in Book XI of *The Confessions* without feeling a kind of temporal claustrophobia. Until he likens the experience of time to the actual time of poetic recitation, there appears to be no room for oneself in time:

> And that one hour passes away in fleeting particles. Whatever of it has flown away is past, whatever remains is future. If any portion of time be conceived which cannot now be divided into even the minutest particles of moments, this only is that which may be called present; which, however, flies so rapidly from future to past, that it cannot be extended by any delay. For if it be extended, it is divided into the past and future; but the present has no space.

From the perspective of the point-present, we cannot measure time because there is nothing to measure. It is not possible to compare the duration of one instant, marked, for example, by the presence of a metrically short syllable to another, marked by a metrically long syllable. One cannot speak two syllables at once as one holds a measuring rod to a length.

We can know time and measure it only in its fullness, when we recognize past and future are present in memory and expectation. The measure cannot be linear but must account for different dimensions of the present, which is not a simple atom but a complex event indirectly implicated in all time, past, present, and future. Only if time is taken in the most radically subjective sense does it manifest an objectivity. Augustine's paradigm for temporal measure is the reciting of a psalm: time has both duration and content. And, of course, in Latin verse, which is measured by syllablic duration, not by a normative count of syllabic stress like modern English verse, time is real rather than normative. Augustine says:

> I am about to repeat a psalm that I know. Before I begin, my attention is extended to the whole; but when I have begun, as much of it as becomes past by my saying it is extended in my memory; and the life of this action of mine is extended both ways between my memory, on account of what I have repeated, and my expectation, on account

of what I am about to repeat; yet my consideration is present with me, through which that which was future may be carried over so that it may become past. The more this is done and repeated, by so much (expectation being shortened) the memory is enlarged until the whole expectation be exhausted, when that whole action being ended shall have passed into memory. And what takes place in the entire psalm, takes place also in each individual part of it, and in each individual syllable: this holds in the longer action, of which that psalm is perchance a portion; the same holds in the whole life of man, of which all the actions of man are parts; the same holds in the whole age of the sons of men, of which all the lives of men are parts.

Time then is an intensive vector, not a mechanical duration. For Augustine, speech and real duration are isomorphic. Cartesian grammatology, on the other hand, is atemporal or proposes the fiction of atemporality. The especially prepared text of a formal proof consists of enumerated steps in deductive reasoning. The Cartesian image is a chain: "... we cannot with one single gaze distinguish all the links of a lengthy chain, yet if we have seen the connection of each with its neighbor, we shall be entitled to say that we have seen how the first is connected with the last." The difference between the Cartesian link and the Augustinian syllable as metrical units is that the links may be complexly concatenated arguments in themselves; entire language worlds can be enfolded into a single link. The experience of temporality in the language of a formal proof is utterly unlike the experience of poetic time. Cognitive time does not answer to *any* externality; it is a psychological dimension. Although it would be left to Kant to deny the objectivity of space, which is implicit in the circular Cartesian argument, Descartes recognizes cognitivity as the source and end of temporality. In seventeenth-century science, therefore, cause remains objective, until it is shown by David Hume to be merely a semantic strategy for interpreting an atemporal calculus, but purpose becomes private, a matter not of "nature" but of will. It was the province, therefore, not of science but of culture.

In the ideal realm which the Greeks discovered, it was the purpose of the acorn to grow into an oak tree, but modern science deals only with ideal or achieved purposes. In the modern view, acorns are always oak trees as they are also dead and decayed oak trees and soil and nutrition for second-growth forests, and so forth. The cosmos is already finished, according to Newtonian doctrine, a mere playing out of the implications of formulae which are apparent to a divine mind. As limited divinities, we can get the formulae but our information-handling capacity is too limited. We do not have sufficient data to predict the future except in carefully isolated and controlled situations. In the Euclidean-Newtonian cosmos, time is a shadowy dimension. Although Newton himself supposed that it was a divine attribute, prior even to the creation of the material world, Spinoza and Malebranche, who were bolder and more consistent metaphysicians than Newton, saw that it was unnecessary to a fundamental description of the world. The principles of the cosmic mechanism imply and supersede all possible events. Laplace gave this succinct formulation in 1814:

> An intellect which at a given instance knew all the forces acting in nature, and the position of all things of which the world consists—supposing the said intellect were vast enough to subject these data to analysis—would embrace in the same formula the motions of the greatest bodies in the universe and those of the slightest atoms; nothing would be uncertain for it, and the future, like the past would be present to its eyes.

Although it is mistaken to say that time in the Newtonian cosmos is illusory, it is a human dimension, peculiar to finite minds, and important as a teleological framework in an otherwise complete and purposeless world of matter in predetermined motion. After the Newtonian world-picture was psychologized by Kant, the model of classical physics was easily extended to the development of history and the social sciences, and with the emergence of technology, it became increasingly clear that the human world changes profoundly despite the eternal nature of the cosmic principles.

In the abstract verse forms which came to dominate poetry in the sixteenth and seventeenth centuries, absolute time was a dominant implication; any particular performance of a poem was an instance of its potentiality or purpose which must be energized by private motives. In this light, the deeper meaning of Northrop Frye's comment on the relationship between the native four-stress English line and iambic pentameter become clear: if stress and meter neutralize one another, even partially, the voice must be seen as poised between the living performance and the deathly form. The voice *refers* to a form which it cannot adequately express. The pattern allows innumerable possibilities for any given choice in the actual performance. That is, it allows performers of the verse freedom, but only inside a context which insures that they thump out some version of pentameter. The voice becomes a temporal function of the metrical form just as the actual things to which the poem refers becomes a function of mathematical analysis. The poem cannot be performed, as the text "contains" any number of possible performances, it can only be *represented* by a performance. The recursion which we noted in the imagery in Sir Walter Raleigh's poem, making the image life into an infinite regression of plays within plays, appears also as a factor in the performance of iambic-pentameter poems; the performance refers to its mysterious origin, not the actual time and space.

4. Probability Theory

If we consider the most dramatic intellectual development of the seventeenth century, the sudden appearance of mathematical analysis and the absolute systems which its intellectual implementation required, the break with the medieval tradition seems clean. Such was the experience of many at the time ("The new philosophy calls all in doubt," etc.), and that break has subsequently provided an essential landmark for our historical orientation. Increasingly, however, scholarship demonstrates that the scientific "revolution" consisted of the appearance of a few new techniques which led to the gradual transformation of traditional modes of thought. As Ian Hacking notes in *The Emergence of Probability: A Philosophical Study of Early Ideas about Probability, Induction, and Statistical Inference,* the doctrine of signatures, that curious, central doctrine of medieval science, was not displaced but transformed:

> It would be amazing if a Paracelsus were an 'influence' on a Pascal or a Leibniz. The mathematicians despised what they knew of the occult. Yet their contempt for those hermetical figures does not preclude the possibility that whatever these geometers thought about opinion, they thought in a conceptual space that was the legacy of the very empirics whom they scorned.

And this is precisely the thesis which Hacking argues. He writes:

> The old medieval probability was a matter of opinion. An opinion was probable if it was approved by ancient authority, or at least was well testified to. This medieval concept of probability is indeed related to our own, but in a surprising way. A new kind of testimony was accepted: the testimony of nature which, like any authority, was to be read.

It is only a step from the natural signature to the concepts of evidence and inductive probability. What is retained from the prognosticating legacy of the occult traditions is a sense of the reliable relationship between an observed past and an unobserved future. Propositions about the conjunctions of past and future, cause and effect, are frequently tenuous and, as David Hume would show, can never be asserted with confidence as they might be in a mathematical proof. The fact remains that some assessments of the signs are more reliable than others, and in the seventeenth century there arose a mathematical theory for describing these statistical regularities.

The first practical applications for this new science, apart from its obvious use in gambling, were actuarial. In attempts to keep track of the plagues, the city officials of London had collected considerable statistical information, beginning in 1603, and by the middle of the seventeenth century statistics were being kept in many parts of Europe. Basing their studies on this material, John Graunt and William Petty established the fundamentals of what Graunt called "political arithmetic." Their work must be reckoned as the beginnings of statistical social science. Its applications were immediate and practical. It was possible, thereafter, to rationalize annuity and life insurance schemes and a little more than a century later, with Thomas Malthus, to argue political policy in statistical terms.

Ultimately, the idealized space of mathematics and the idealized subject of poetry were necessary only to initiate progressive institutions. Algebraic space and the modern ego were the first products of the new technology, and they have been rendered obsolete by their own accomplishments. If we were to emphasize the continuity with the seventeenth century, we could say that Leibniz and Pascal, who were both deeply involved in the develop of probability theory, rather than Descartes, have emerged as the most important instigators of our thought, or we can say that Descartes' second house of culture has appropriated the necessary mathematical means to be technologically self-sufficient. Therefore, separate ideal realms for the mathematical object and the human subject are no longer required, and both mathematics as a form of disinterested meditation and poetry have largely decayed. The logic of statistics replaces the logic of the absolute; that is, a logic of ratios between determinate events and definitive possibilities replaces the logic of ratios between determinate events and infinite possibilities.

Our ultimate concern is to note the erosion of the place of the ideal subject of which poetry was the form and to establish the sociocultural context in which poetry must now be produced and read. The culture developed in relation to its most idealistic conceptions because they were clearest and most powerful. Having reached a point of strength, it is now possible as it were, thinking of Wittgenstein's metaphor, to kick away the ladder which got us here. Of course, knowledge did not become suddenly statistical when an articulate theory of probability finally emergenced. It was necessary for the idealist tradition somehow to account for those aspects of reality which are now described statistically:

The distinction between the identity, *I am this one that is one*, and the identity, *I am this all that is all*, is conventional rather than logical. We do not know how large one is. We speak of *one* person or *one* species or one cosmos. That is, we have a problem of measurement. Leibniz intuited this problem, and, though he did not develop the mathematics of it, his monadology consciously embraces the contradiction: a monad is both atomic entity and cosmos, an individual and everything. However, he notes precisely the limits of the representation; there is an inevitable confusion or blurring of the image as it reaches toward infinity:

> ...things cannot be otherwise than they are. It is because God, in ordering the whole, has had regard to every part and in particular each monad; and since the monad is by its very nature representative, nothing can limit it to represent merely a part of things. It is nevertheless true that this representation is, as regards the details of the whole universe, only a confused representation, and is distinct only as regards a small part of them, that is to say, as regards those things which are nearest or greatest in relation to each monad. ... In a confused way they [the monads] reach out to infinity or to the whole, but are limited and differentiated in the degree of their distinct perceptions.

This states the nature of the statistical subject: it is able to define local regions with considerable clearness and distinctness, but with respect to the whole, its picture is necessarily fuzzy and its apprehension of itself is based on certain zones of statistically-reliable frequencies. This example alone should allow us to recognize that the historical processes are not neat narrative progressions. In the generation after Descartes, this alternative to the clear and distinct subject was forcefully articulated, but it had to wait until this century for a practical mathematics to catch up with it. The nature of the equivocation was clearer in the calculus. Bishop Berkeley noted certain problems in mathematical analysis on which the whole of Newtonian physics depended. In its standard seventeenth-century development, the calculus required an undisguised double shuffle: it dealt with infinitely small quantities which sometimes counted as significant and sometimes did not. In the standard calculus text of the early eighteenth century, the first postulate declared, in effect, that "a differential can increase a quantity without increasing it." In "The Analyst; or a Discourse Addressed to an Infidel Mathematician Wherein It Is Examined Whether the Object, Principles, and Inferences of the Modern Analysis are More Distinctly Conceived, or More Evidently Deduced, than Religious Mysteries and Points of Faith" (1734), Berkeley asks, "And what are these Fluxions? The Velocities of evanescent Increments? And what are these evanescent increments. They are neither finite Quantities, nor Quantities infinitely small nor yet nothing. May we not call them the Ghosts of departed Quantities?"

The mathematicians were fully aware of the contradiction, but the postulate was necessary to results which were demonstrably useful and important. The relevant metaphysical question involved the existence of infinitely small quanitities. Leibniz, consistently the most scrupulous of the great seventeenth-century thinkers, doubted the existence of such quantities and took a formalistic stance, arguing that, although the symbols of the equations in themselves are meaningless, their use allows one to determine important results. In time, mathematicians began to emphasize not the variables, which change in infinitely small increments, but the functions which determine them, thus at least partially hiding these "Ghosts of departed Quantities" from theoretical view. Even as late as 1904, however, Gottleb Frege writes:

> It is even now not beyond all doubt what the word 'function' stands for in Analysis, although it has been in continual use for a long time. In definitions, we find two expressions constantly recurring, sometimes in combination and sometimes separately: 'mathematical expression' and 'variable.' We also notice a fluctuating usage: the name 'function' is given sometimes to what determines the mode of dependence, or perhaps to the mode dependence itself, and sometimes to the dependent variable.
>
> In recent times the word 'variable' is predominant in the definitions. But this is itself very much in need of explanation. Any variation occurs in time. Consequently Analysis would have to deal with a process in time, since it takes variables into consideration. But

Symbolic Nature

in fact it has nothing to do with time; its applicability to occurences in time is irrelevant. There are also applications of Analysis to geometry; and here time is left quite out of account.

It requires little mathematical sophistication to appreciate the problem here. If one emphasizes the abstract nature of the function, the relationship appears as atemporal. For every y there is a corresponding x. Nothing about this relationship is variable; it is always true of all y's and all x's. However, in scientific applications, one is likely to be interested in the relationship between a particular y and a particular x. What, for example, is the trajectory of a cannon ball if one varies the charge? and so forth. We require some kind of metaphysical hoo-doo to allow the simultaneous consideration of particularity and variability. The scientific project has invested considerable energy in hiding its metaphysical traces.

The solution to this problem involves the recognition that X is P not in relation to an absolute context but in relationship to a presupposed and definable environment: that X is P with respect to R. The reconcilliation between the ideal subject which can cognize the abstract relationship X is P and nature in which X is more or less P requires a mathematical theory of statistical stability. The ideal subject cognizes all possibilities and is, therefore, conceived as a descriptive medium. Judgment or the attainment of knowledge arises from the determination that certain possibilities are realized with reliable frequency. Knowledge is expressed as a ratio between enumerated possibilities and actual occurrences: X is $P, Q, R, S,$ or T and X is P. If the the possibilities cannot be specified, then neither the purpose of the judgment nor the value of the proposition can be assessed. During the past century, knowledge, which had been conceived as the ratio of propositions to things, was reconceived as a ratio of possibilities to particular occurrences. Statistical reality subordinates the individual not to a *logos* but to a relevant population of humans, electrons, or whatever. The correlation between the system and the corresponding process is statistical; it does not apply to any individual.

The interior dimension of human understanding, where purpose is determined and acted upon, was beyond the horizon of the early modern *episteme*. It failed to account for its implementation of itself which is its most apparent and useful character. By the middle of the nineteenth century, the interior of the subject was articulated as possibility. The statistical subject is the Cartesian subject conceived not as a clear and distinct unity but as the repository of possibility, which is to say, the Freudian unconscious or something like it. It was however necessary to sentimentalize poetry and the arts in general in an attempt to tame the wild individualism from which its formal possibility sprang. In this century, poetry's function has been transferred to the quasi-formal languages of the social sciences. Thus, we have been overwhelmed by the objectivity of language.

On the Logic of the Moral Sciences

A.S. Yessenin-Volpin

Translated from the Russian by
Elaine Ulman, Karen Andreason,
and Christer Hennix (ed.)

Foreword

This treatise, written in December of 1970, is the second of three philosophical treatises which I decided to write when my research on the development of a concept of proof in general logic reached a stage where it could be interrupted. The first of these treatises, entitled "On the Antitraditional (Ultra-Intuitionistic) Program for the Foundations of Mathematics and the Natural Sciences," was written a few weeks earlier. It does not have any direct relation to generally recognized problems, and I mention it here only to note the place these problems occupy in the system of scientific studies I recommend. In this system, the primary role should be allotted to the struggle against the necessity of faith and the development, as far as possible, of universal and irreproachable methods of proof. Until it is completed, the basic focus of my research will be related to the foundations of mathematics. But in that very sphere I clarify a close connection with the principles of the logic of the moral sciences, principles which must always guide investigations of the deepest questions relating to the nature of rules and goals ('tseli') and the understanding of truth and evidence. I intend to devote

---

Originally published in *Social Problems* No. 12, 1972 (in Russian).

the third of the aforementioned treatises to purely philosophical questions.

In a scientific sense, this treatise does not pretend fully to clarify all the logical principles which are of interest in connection with its theme and which I discovered in the course of developing my antitraditional investigation of the foundations of mathematics. In particular, the semiotic principles used in my investigation have been omitted. The question of limits to the applicability of logic has also been completely set aside. In my basic research this question is connected with expressiveness in language which, in the case of processes, I identify with their discreteness. I mention this in the foreword in order to avoid reproaches of trying to implant [my ideas] too singlemindedly, without taking the limitations of rationalism into account. But in so mentioning the problems of the limitations of language and possibly, in connection with this, of logic also, I have no intention of advocating without proof any theses which assert this limitedness. Logic has no need of such theses in any case, and I at least attempt to prevent such theses from posing a threat. For this reason logic now needs to be expanded and deepened in every possible way. The foundations of the moral sciences ought to be grounded in a rationally-based logic so that one might always demand a total explanation of any doctrine suggested by these sciences and the motives for its acceptance.

In this treatise the important terms 'coercion' and 'fraud,' and perhaps several others as well, have been left without a definition or thorough explanation. I shall not try to remedy this omission in the present foreword, but in time I hope to return to this theme in another work.

Currently I am writing the second part of my work [3] in which I particularly use the concepts of permission, basis, and fundamentary regime herein explained.

24 August 1971

On the Logic of the Moral Sciences

I shall discuss the logic of ethics and jurisprudence. Up to the present time, certainly, much has been alogical in these sciences. They developed mainly as an expression of historical and political tendencies and the standards of acting legislators. If one where to search for the logic of the moral sciences in their process of development, then one would enrich one branch of logic above all others—that of logical errors. However, logically an alternate course of development of these sciences is conceivable wherein moral and juridical systems of rules or norms are established in strict correspondence with propositions of an impartially developed logic. Such a course is certainly not historical, if one speaks only of the history of the past and present, but its study might influence the history of the future, especially if thinking people reject the harmful habit of elevating the ruling lawlessness and alogism into law. Thus the study of logic and its connection with the moral sciences produces important preconditions for moral progress though it must be understood that the factual realization of progress demands not only abstract theorizing, but also ceaseless struggle for enlightenment, struggle with many dangerous human vices. (I place deceitfulness first among these because it serves as a screen for other vices.) Doubtless this struggle will go on for a very long time before significant, observable successes occur in the practical realm. But however the matter stands in relation to this struggle, the development and expansion of the logic of the moral sciences ought to be an absolutely necessary condition of the struggle.

In this brief essay I can cover only a few questions from the areas of logic being examined, and then only briefly. These branches of logic have not yet been widely developed, and in my opinion the time for writing an extensive work encompassing this entire area of logic has not yet arrived. I say this consciously disregarding the numerous articles and essays written on these themes—they have not yet achieved a unified logical approach.

I examine the close connection of the logic of the moral sciences with problems in the foundations of mathematics as presented in my works [1–3]. In these works I developed an analysis of the difficulties in the foundations of mathematics to the point where the theory of modality and other "prototheories," i.e., theories preceding the elaboration of methods of logical proof, were included. The fundamental rôle of rules was demonstrated, and an original logic of rules evolved. Clearly, rules play a fundamental rôle in logic and mathematics as they do in ethics, jurisprudence, and also semiotics (including linguistics) and psychology as well. This leads to the connection between the aforementioned branches of logic and the foundations of mathematics, a connection deepened by the need to examine the most important principles of *preference, collation* (i.e., identification and discerning), and *acts of attention* (i.e., *following* and *neglecting* connections) in the foundations of mathematics with such generality that the (analysis) is independent of the subject of mathematics and extends to any science, including the moral sciences.

I shall begin the discussion of these questions with the division of all linguistic propositions into the following classes: *rules* (i.e., *permissions* and *demands,* including *proscriptions*), *goals* ('tseli'), *desires, judgments* (i.e., a statement A, for which the question, "Is A correct?" is possible), *requests* and *commands* (including *questions,* considered as requests for an answer), and *names of actions and events.* This classification does not pretend to be complete, but I do not foresee the need to discuss propositions not included in it.

Logic is the science of standards of correct reasoning, the study of avoiding errors. In all fields of human activity where the risk of error is recognized as intolerable, the rigorous use of proof is demanded. I call each occasion when a judgment is recognized as true without proof *faith.* Faith is always connected with the risk of error and this risk continues so long as "truth" remains accepted without an explanation which would constitute proof, without an answer to the question, "Why is this accepted?" In the area of acceptance of judgments, the *law of sufficient reason* consists in considering *only* proven results as true. By proof of a judgment I mean any honest method which makes the judgment incontestable. A theory of disputes, in which this understanding of incontestability is more precisely defined, is needed; by *honesty* I mean the absence of *coercion* and *fraud,* concepts which I am prepared to make explicit in the prototheories. In the area of acceptance of rules, the *law of sufficient reason* consists in the demand that the *understanding* of every rule be *grounded.* I will refrain from the needed elaboration of what has been said for now, only noting that the elaboration will be connected with the nature of rules or judgments similar to these, regardless of the areas to which they are related. For this reason it will proceed in a similar fashion in the foundations of mathematics and in the moral sciences.

The prime goal I see in the foundations of any science is the complete banishment of faith. Perhaps faith is necessary in various spheres of human activity, and in any case the right to have faith constitutes an inalienable part of freedom of thought, but there is no freedom where this right places a thinker under obligation. Everyone must have the unlimited right to ask "Why?", the question that destroys any faith. Therefore, if acceptance of a judgment is required, proof of that judgment must be shown. True, proof does not create a requirement that the judgment be accepted; according to the law of sufficient reason it creates only the right to accept. Generally speaking, the obligation to accept a truth, i.e., the results of a proof, is required only for the achievement of some accepted goals or the fulfillment of some desire. The right to doubt proven judgments is distinct from the right to criticize a proof, which in practice is more important. For the skeptic, a proof offers a basis for accepting the proven judgment, a basis which he may or may not use. But in all cases when adopted goals or desires make it important to accept a judgment, such a basis is taken to be satisfactory, and the acceptance of a judgment on the basis of proof, in any case, does not appear as faith.

In the history of human thought the need for faith was called forth by a weakness in the ability to reason and argue, a weakness that can be overcome only by expanding and deepening logic and widely disseminating the information acquired in this way. To this day the inevitability of faith remains a commonplace conviction among the majority of thinking people. So long as logic is limited to a few branches, as has been the case until now, this conviction will be reinforced by widely recognized arguments. The arguments insist upon the need for and place a very high value upon faith, and even the word itself, "faith," acquires an expanded meaning. "Faith" is often understood as preferring to accept certain judgments rather than reject them regardless of any rational arguments, irrespective of the presence or absence of proofs, even holding to the preference with particular stubborness. Faith is demanded as a necessary condition for the continuation of joint activity among people, and it must be admitted that such demands have a practical use. But they limit freedom of doubt as well as freedom of criticism and freedom of thought in general. In itself faith does not diminish freedom of thought; it even seems one of the manifestations of this freedom and can have non-negative value. But demands and coercion to have faith limit this freedom in a most essential way, and this applies to faith in any understanding of that word. Thus any moral system, any legal system demanding even the smallest degree of faith limits freedom of thought, and he who accepts them is no longer free in his judgments about the validity of acceptance, and there is no rational basis for having confidence in his judgments. For this very reason the moral sciences must be developed without any recourse to faith. A moral system based on faith is admissible only for followers of that faith, but coercion into a faith is immoral because it limits freedom of thought. Of course, this argument is valid only for those who value this freedom, without which, certainly, no rational basis for trusting the achievements of thought and cherishing moral values exists.

For this very reason I attach fundamental importance to the search for the foundations of the moral sciences, foundations as fully independent of faith as possible. Perhaps, however, the complete banishment of faith may remain an unattained ideal. In that case I recommend the development of a special branch of logic which I call the *logic of confidence* and which I oppose to the "logic of proof." The logic of confidence claims to have full control over applications of faith (from here on I will use this word only to signify the acceptance of judgments without proof). If in the development of the foundations of a science (or any other field of activity) there comes a moment when faith, though not completely banished, remains only at the peripheries, where everyone tolerates it, then that science (or other field) is all the same sufficiently grounded. (Although there is a particular sort of people, "fundamentalists," who consider it their calling to continue criticizing and improving the foundations, and I too ask that their efforts be met with deep respect.)

Although the understanding of proof is connected with the understanding of incontestability and, by the same token, with the theory of disputes, I by no means intend to look upon proof as a procedure necessarily containing within itself the construction of a dispute. On the contrary, it is a method intended to avoid dispute. There is at least one method which everyone is forced to regard as proof—that is the application of a definition or, more generally, any (accepted) rule for the use of signs. For example, the definition, "a bitch is a female dog," strictly speaking, must be examined as a system of two rules: permission to call a female dog a bitch and the demand that the word 'bitch' be used only in accordance with this permission. Using this permission, we derive that a female dog is called a bitch, and then using the definition of the connective "is," in the same way consisting of permission to replace the word "call" by the word "is" and the demand that this word be used in accordance with this permission, we derive the theorem, "a female dog is a bitch." Using the aforementioned demand instead of these permissions the reverse theorem can be derived: "a bitch is a female dog." (I should note that a full proof of these theorems would look more complicated, and the second of them is bound up with profound subtleties in the theory of the use of the word "only.")

Of course, this method of proof can be applied to any definition, regardless of the essential meaning of the concepts being defined. It makes the judgments being proven incontestable if only because, according to the rules of honest dispute, each side must follow the accepted rules (which, by the way, may appear more complicated than the definitions being examined) regulating the use of signs. The art of logic must consist in building any argument according to the concept of some procedure governed by the application of this method. Of course, in complicatd cases the argument is not exhausted by these methods, but other methods should be used only in order to crown the argument by the application of accepted rules of sign usage. Proof of the presence of a table lamp in a room might include, as the most obvious method, pointing out the lamp to the addressee of the argument. But in a logical analysis the argument would not be settled by this method. Not the presence of the lamp, but rather a *judgment* about that presence is proven, and therefore the argument must include an application of the rules of sign usage entering into this judgment: "there is," "a lamp," "in the room," etc. The proof is completed only when all these necessary rules have been applied in the proper way.

I do not wish to insist without proof that the rules of sign usage are the only ones to play this role in proofs, but the necessity of applying any other such rule must be proven in an honest debate (in which case the concept of honesty must make provision for the right of each participant in the debate to carry the argument through to the end and forbid any "obstruction," as impeding the implementation of this rule). But at the present stage of investigation I do not foresee the need for drawing any rules other than those of sign usage into this role. For example, it may be found necessary to clearly specify permission to rely on memory, but I intend to consider such permission one of the rules of sign usage.

In the course of the arguments the following steps are performed:

collation, acts of attention (see above), indications, perceptions, and acceptances of propositions. In very deep examinations of arguments, to these steps must be added acts of preference (or choice), acts of reference to memory, and, when linking the concepts of proof with the theory of disputes, also acts of reference to another person. Denials of and abstentions from these sorts of acts, as well as from the denials themselves, must also be allowed. Proofs are effected in a *theoretical activity.* In the important *theory of reasonings* having to do with criteria of "correctness" of judgments (also consisting of the performance of the aforementioned types of acts), rules are established characterizing one or another theoretical activity. In the light of these criteria (which I will investigate in more detail in the second part [3]), only proven judgments will be considered correct. (As already mentioned, perceptions may also be a part of judgments, but I will not pause for a more detailed account of this here.)

In respect to syntax, the aforementioned logic of confidence has to do with the transitions from the statements "A says B" and "A is correct" (or "what A says deserves confidence,") to statement B. In this case A is called the *source* or *bearer* of confidence. But the most essential part of this logic consists of the principles on which the choice of a source of confidence is based. In particular, authorities, witnesses or experts, books, material evidence, branches of science, mental faculties (i.e., memory), etc., may serve in this capacity. Each source (bearer) of confidence has his *sphere of competence,* and confidence, always proceeding from some person, must be based on a particular *act of confidence* which can be rejected. The logic of proofs, as well as proofs of the correctness of assertions from a certain source which are related to a certain area must be considered the most perfect sources of confidence. In all cases rejection of any act of confidence is permitted as long as the rules of the theoretical activity do not forbid such rejection. The rejection of all testimonies from the source of confidence (i.e., of all propositions B asserted by the source) and of all other propositions accepted through the agency of these testimonies must follow as a consequence of this rejection. However, a restriction of the area of competence may occur in place of a rejection of an act of confidence. This may be thought of as a rejection of an act of confidence accompanied by a new act of confidence toward the same source but with a restricted area of competence.

The following may serve as *grounds for the rejection of an act of confidence:* a) Errors on the part of the source of confidence not considered possible at the time of the act; b) Deterioration of conditions for verifying the testimony of the source compared to what is expected during the act of confidence; c) Behavior on the part of the source of confidence which would promote such deterioration (in particular, violations of the principle of publicity (*glasnost*) by the courts, etc.); d) A conscious lie authorized by the source of confidence in testimonies or sometimes even in judgments not related to his sphere of competence; e) Well-founded speculations about the deterioration of the source of confidence's capabilities or honesty; f) Discovery of a better source of confidence; g) Increase in the demand for truthfulness, precision or authenticity in the testimony of the source of confidence; h) Resemblance to another source which has been denied confidence. This list of grounds is not exhaustive.

The search for such grounds is called *criticism of confidence* and evolves in accordance with the methodology of investigation in a given activity. The methodology may demand various specifications, for example, indication of occasions when a conscious lie outside the sphere of competence is considered grounds for rejecting an act of confidence (applicable to point e), etc.). In the case of point c), (and also in the case of a decline in honesty) the source of confidence becomes "suspect" and, in the case of d), a "liar," which affects an "evaluation" of it, that is, the preference of other sources to it over sources resembling it.

In any case the persuasiveness of the testimony of a source of confidence cannot surpass the persuasiveness it possesses for the source itself—this means particularly that the testimony must be interpreted with all reservations and doubts that the source has or ought to have. On the other hand, if the falsehood of the source of confidence reveals a system which allows one to apply a way of correcting the testimony, an act of confidence to a new source is possible rising from the former act by appling these corrections (as one does with a clock, for example, when one knows how many minutes slow it is, etc.). In the case of point a), revelation of the cause of error may be considered grounds for a continuation of confidence in those cases where this cause cannot operate (representing only a slight limitation of the sphere of competence), but it may also demand the elimination of this and/or similar causes. The evaluation of the source of confidence can also depend on the range of such causes. Knowledge that such causes can operate only rarely should entail an improvement in the evaluation, and this argument can be applied to the selection of a source of confidence.

In general, grounds for an act of confidence must be connected to the competence of the source. In some cases a "tautological connection" may successfully be established—for example, our sensory organs are the best sources of confidence for evidence about actual reality in as much as actual reality is considered the contents of their evidence. In many cases the methodology of an investigation allows one to trust one source more than another similar one which has sufficient reliability but is inaccessible. The absence of substantial grounds for differentiation in their evaluations must serve as a criterion of similarity in this case.

'Trial acts of confidence,' made without any grounds, are possible in order to see what transpires. The *heuristic principle of confidence* plays a most significant role in human cognition: if, notwithstanding the application of all available methods of criticism of confidence, there are no grounds for denying confidence to a source, an act of confidence is made toward it. The evaluation of this act must depend on the completeness of the aforementioned methods and improves in proportion to their reenforcement. Applications of induction through simple enumeration and of the phenomenological principle in the natural sciences (consisting of a theory's being wholly or partly accepted when all its assertions have been confirmed by observation) are based on this principle. Acts of confidence toward sources which have undergone a trial act and have not been deceitful are also based on this principle.

In the moral sciences the theory of modalities, especially deontic and optative modalities, must play a no less central role than it plays in the foundations of mathematics. I shall describe this theory only briefly here; its contents have been presented in more detail in my work (1)-(3).

Modalities are divided into three categories (possible, actual, and necessary) and five groups: deontic ('possible' means 'permitted' or 'allowed'; 'necessary' means 'required'), optative, that is, connected with goals or desires (in relation to the achievement or realization of which 'possiblity', 'necessity', etc. are discussed), and three alethic groups: organical (connected with means, including ways — 'possibility' means knowing how or having the ability to perform an act under consideration, 'necessity' means that one is compelled by an accepted manner of acting or the character of the process under consideration), epistemic (connected with the cognitive process — the possibility of an event means the organic *(organicheskaya)* possibility of continuing that process, assuming that the event will occur; the necessity of an event means the organic necessity of accepting the assumption that it will occur), and ontological (connected with the reality being examined; if the latter is a process defined by some means, it is simply the organic modalities connected with these means). There are further distinctions among these groups, particularly those related to single and double negations.* The zero-modality "actually" exists for all groups; those assumptions and judgments expressing perceptions and opinions adopted in the course of or at the basis of an examination are accepted with epistemic actuality.

Modalities are applied to judgments or propositions designating acts (actions or inactions) or events. More precisely, types of these propositions are indicated in a natural way for each group once and for all (deontic and organic are applied to acts, epistemic to judgments, etc.). These propositions aside, modalities are always related to circumstances (what is possible in some circumstances may not be possible in others, and analogously for all categories and groups of modalities). In addition, the same modalities apply to circumstances. Judgments are the result of the applications of modalities to propositions under certain circumstances, except in the case of the deontic modalities "it is permitted" and "it is required" when such results are rules. Demands to refrain from an act are called *prohibitions* or *bans* of the act.

The circumstances must be indicated in some way or other. The common way of indicating circumstances consists of describing them by means of some set or class of judgments — in this case I call the circumstances a *situation*. To be exact, a *situation* is the description of the circumstances by judgments and may be more or less precise, but when a sufficiently precise description is present, the situation is identified with the circumstances. In abstract reasonings and in the formulation of rules, circumstances are simply presented as situations and therefore are identified with them.

Situations on which modal propositions have bearing (those in which modalities have bearing on other propositions, but not on the circumstances) can depend on parameters, and be therefore, indeterminate and can themselves serve as parameters for situations (representing classes of meanings for these indeterminate situations). Situations can be designated in a list by letters and indices, but such indices

---

* The "necessity" splits in two modalities: "compulsoriness" (or "obligatoriness") and the "necessity", i.e. indispensability or "impossibility" of negation (or of opposed event). Here, for simplicity, this distinction is smothered and mostly suppressed.

can degenerate in accordance with general rules of the usage of indices, particularly when their meanings are fixed in a context or do not play a rôle.

The following *principle of modal fulfillment pmf* has a variation for each group of modalities: If situation S is possible, and A is possible in S, then situation S + {A} is possible, resulting in the addition of judgment A to S.

More precisely, if in *pmf* one discusses any possibilities other than epistemic and ontological ones, then A is the name of an act or event, and, in the proposition A is possible in S, the given A must stand in the future tense or in the infinitive. But in S + {A}, A must stand in the past indicative. If the possibility relates to performing act A by an agent, then in the first case A is used in the active voice, but in the second, in the passive. In the case of both epistemic and ontological possibility, A can stand in the present indicative, but in the first case it can stand in the future tense and in the second in the past. In the statement "A is possible in S" as soon as A is in the future tense, according to *pmf*, the situation S + {A} is considered possible in (the same) future (which contains some specification of *pmf* for the temporal status of a situation). In any case, in *pmf* all three possibilities must belong to one and the same group.

Most significant in *pmf* is the transformation of A from possible in S to actual in S + {A}. This transformation is accomplished at the price of the situation's being considered only possible, even if S were actual. This constitutes the fullfilment of the modality for A.

The *pmf* is the principle by means of which applications or realizations of possibilities are formalized. Unlike the rules of deductions, S + {A} is considered only possible. This is the rule of the addition of a new judgment (or assumption) to the situation being considered, realized within the bounds of possible situations.

Situations are called *actual* if all their elements are accepted on the basis of perception; such actuality is called *real*. *Speculative* actualities, some elements of which may be accepted as hypotheses or consequences of other accepted propositions according to the rules of logic adopted, are also examined. (In many instances these rules, as well as the acceptance of some hypotheses, are assumed to be inherent in the subject under consideration and therefore need not be specially stipulated. In such cases actuality is considered real despite the usage of these rules and hypotheses.)

Actual situations are considered alethically possible (for any alethic group). The grounding of *pmf* can be that only those situations which can be confirmed as possible on the basis of *pmf* are recognized as subsequently possible. Such a grounding is tautological and therefore incontestable. To apply it to any group, it suffices to agree beforehand on which situations will be considered actual and which judgments of the form "A is possible in S" (representing "the possibility of S + {A} in relation to S") will be accepted in the theory or activity under review.

I call a system of rules a *character* or *tactic*. The first of these terms is chosen in accordance with the fact that the character of any subject is considered known as soon as the rules governing how it may and must act in every situation are known. Under this 'act' may be included everything that can be expressed by a statement with the verb in the infinitive, in particular, to think about or say anything whatever, to smile, to forget, or to select a new character. Thus, the concept of character embraces any case of a regular change of character, and connected with this is a certain natural hierarchy of characters. But the very same system of rules defines the character not only of an agent, but also of his activity; I call the latter the 'tactic' of the agent. Generally only those situations and actions which are not completely arbitrary and which one may meet during an activity or process under consideration and may need to discuss in this connection, arouse interest and reveal character. Hence two important classes arise—the *class of situations* and the *class of acts*—in relation to every character or tactic examined. If a character (or tactic) is expressed in language, I call it a *method*. Aside from the two classes mentioned above, some basis to aid in reaching this expression always is crucial for a method. When the distinctions among the concepts of character, tactic, and method do not play a role, I use the word *way* in the same sense.

In a situation S, to *follow* the rule permitting or requiring one to perform act A *in* S means to perform A in S. Only a requirement can be *violated,* and *breaking* the requirement A in S consists of performing the *opposite* act in S (i.e., not-A, or B, if A is not-B). To *follow* a way in a situation S means to follow every requirement of this way in S and, in the absence of requirements related to S, to follow at least one permission related to S. To *follow* a way in all situations of some *class* means to follow it in any situation of that class. Generally one speaks about *following* a way beginning with some given situation (or several given situations) and continuing in all situations arising as a result of the performance of the steps of this application. I call the following of a method a *discrete process* or discrete activity. (If one speaks of process, then in the previous discussions "acts" must be replaced by "events" which will be considered "events of that process.") I say that the method *describes* that process. (In place of method one may sometimes speak of a tactic in the same sense, describing a process.)

Every discrete process is generally performed against a background of other processes *external* to or *deeper* than the process itself. The following may serve as examples of external processes related to ordinary, not very deep, theoretical activity: the collation of judgments or parts thereof, or acts of attention to such judgments or, let us say, acts of pronouncing separate words. Such external processes appear as activities in their turn. Events of external processes not belonging to a given process are performed *automatically* or "by themselves." During the study of a given process external processes are attended to only gradually, as they are needed. Generally some external processes are considered well known, which gives the possibility of conducting investigations related to these processes much the same as one would with the "initial" ones.

In the moral sciences rules are often examined in more detail than is usual in the foundations of mathematics (if one disregards the theory of disputes). To wit, in the moral sciences an *addressee* is indicated, i.e., a person (or persons) who must follow the rules, as well as an *addressor,* i.e., a person by whose will the rules are accepted. Only the addressees of rules must follow them, and only they may violate them. But this is not a unique characteristic of the moral sciences, and if these persons are not mentioned in logical and mathematical theories, it is only because they are identified with the reader or other addressees of these theories.

Ethical systems defining rules of conduct or jurisprudence may serve as examples of methods on the one hand; on the other hand games, fully constituted grammars, mathematical algorithms are also examples.

Deontic judgments, that is, judgments about a rule's being accepted, are associated with rules. Thus, a notice reading "No smoking here" is understood not as a rule, but as a judgment about the acceptance of a rule (in which the word "here" designates the situation). Not being judgments, rules cannot be true or false, cannot be an object of faith or follow one from another according to the rules of logic. This is possible for deontic judgments, but logic does not have general rules by which deontic judgments could follow one from another. In the general case, at least, one can examine utterly preposterous systems of rules.

In external form, deontic judgments and rules can be expressed by the same words in a natural language. In such cases the character of the proposition has to be recognized from its context. To this end, when necessary, special explanations can of course be introduced into the text. In specialized languages the appropriate signs can be systematically used instead.

In natural language modalities are often used together, with one proposition being understood in various senses corresponding to the various groups of a single category of modalities. This often leads to double meanings which are impermissible when absolute precision is demanded, but such precision may also be achieved with the help of philological explanations and stipulations.

In formulating the rules of a method, one may permit oneself simultaneously to permit and prohibit the same act A for the same situation S. This does not prevent our applying such a method, but, according to the description of this application, permission for A will turn out to be inapplicable in S, in which case one must follow the prohibition (for it then represents a demand). It must be noted that in jurisprudence one runs across such *clashes* between permissions and prohibitions fairly often, in fact it would be difficult to avoid them, but on the strength of the previous discussion this need not be so. But one must distinguish such cases of permissions from the rest. For this reason I will call the *allowance* or *authorization* of A in S (from the method side) the presence (in that method) of permission for A in S in the absence of the prohibition of A in S. (I will consider this concept applicable to other ways *(sposob)* as well.)

A *deontic judgment* often represents a permission. Generally in examining deontic modalities permission itself must be considered as 'deontic possibility' or 'authorization' or 'legality', and *pmf* must be applied to it. In this case only 'authorized', not simply 'permitted', situations are considered 'legal'. One can, however, consider even simple permissions as deontic possibility, but in such a case, when permissions clash with prohibitions, deontic possibility may turn out to be deontic impossibility. This will represent a dangerous confusion, but not a real contradiction or absurdity since the phenomenon is well explained by the facts of clashes discussed above. This confusion may be avoided by distinguishing the impossibility of A from the obligatoriness of not-A in the sphere of deontic modalities; equivalent identifications for any group of modalities demand a grounding which operates within broad limits. A certain awkwardness in such distinctions obstructs the freedom of allowing clashes between permissions and prohibitions, but this argument falls away as soon as deontic possibility is considered as authorization. It is certainly not always important to be concerned about this subtlety, and 'authorizations' are often loosely spoken of as 'permissions'.

A method (or any other way) may be 'incomplete' in two fundamentally different respects. On the one hand, for a single situation S it may contain several fundamentally different allowances, for acts A, B..., without making any choice among them. I call such situations S *Buridanian*. A difficulty arises in applying a method in a Buridanian situation, especially if more than one of these acts is feasible in the situation. One way of overcoming this difficulty is to perform all the acts, but this is not always feasible: they may obstruct each other in S. Another way is by examining every variant, but this may not reveal a choice among them. The process of developing a method in this case splits into several equally justified processes (each of which is described by the method). No single one of the processes can be counted as originally feasible in S until a way of choosing a variant has been shown. Finally, a way of preferring one of these acts may be shown (it may be contained in the description of an external process), and then the Buridanian situation has been overcome.

Ways of preference (or 'tastes') play an important rôle all in connection with overcoming Buridanian situations. In addition, they play a most important rôle in the selection of goals or means for reaching them.

The absence of any rule for A in S is another case of an incomplete method. In such cases completion usually occurs with the help of one of two preferences: either permission is preferred to prohibition, which constitutes the *principle of liberalism,* or, on the contrary, prohibition is preferred to permission, which constitutes the *principle of despotism.* It is simplest to apply one of these principles consistently in connection with the rules discussed below. This leads to the application of one of two *regimes:* everything not prohibited by the method is permitted by the regime (*'liberal regime'*), or everything not permitted by the method is prohibited by the regime (*'despotic regime'*). In the first case the distinction between permissions and authorizations does not play any rôle (that is, permission automatically leads to authorization by the regime, but in the second case what is not permitted must be specified (since what is permitted, but not authorized has already been prohibited by the method).

The often encountered confusion of the concepts 'not permitted' and 'forbidden', or 'not forbidden' and 'permitted', is based on the assumption that every act is either permitted or forbidden — and then only one of the two. But only 'complete' methods, almost never encountered (in complex cases), satisfy this assumption. For these methods, actually, the two opposite principles or regimes would be of equal value and would not lead to a noticeable broadening of the system of rules. For incomplete methods, even replenishment by the imposition of one of these regimes does not necessarily lead to completion. (Thus, assume that as a result of the act of imposing a despotic regime over what is not forbidden, a foundation is given only for the negation of the fact that something is not permitted. From this one cannot conclude that the thing is permitted without eliminating the double negation.)

Generally speaking, in actuality three sorts of situations can be distinguished: *creative* — those in which means for attaining goals are sought; *control* — those in which proposed means are verified; and *executive* — those in which the selected means are applied. In the course of an activity they may alternate with each other, forming *stages,* each one consisting only of situations of the appropriate sort. Like prohibitions, permissions are inherent in each of these stages. But in the creative stage one must consistently be guided by principles of liberalism since the opposite preference would limit freedom of inquiry (needlessly, since necessary limitations must be provided for by the method, not by the regime). In the executive stage one must be guided consistently by the principle of despotism (under the threat of not attaining the goal by the means adopted for this). Every verification, in as much as it represents the application of previously adopted means of verifiation, belongs to the executive stage of verifying activity, and therefore in the control stages one must follow the principle of despotism consistently. I count these rules to the number of most significant rules for every methodology.

Generally one considers only those methods in which everything that is demanded is at the same time permitted. In a method that can be effectively applied, what is demanded cannot be forbidden.

The aforementioned principles and regimes are not the only ones imaginable. But there is one reason that makes forming systems of rules in terms of permissions and prohibitions, only rarely including other demands, more convenient. The fact is that these directives generally relate to actions, not abstentions from actions. Various feasible actions may be incompatible, but all inactions, as a rule, can be considered compatible but all inactions, as a rule, can be considered compatible. As soon as a method demands several actions for situation S, for the fulfillment of its development in S one must be concerned about the compatibility of these actions. But these concerns fall away as soon as all these demands, except, perhaps, one, turn out to be prohibited actions.

For any activity and any of its situations S, authorization of A in S, "contained" in the rules of the activity, is the *basis* for performing acts A in S. More precisely, this is *the primary basis*, for, in addition, everything upon which the application of the primary bases is based is also called the *basis* (in the figurative sense) — rules and circumstances of external processes used in the application as well as circumstances characterizing the given situation S. I call the demand that every phase of an activity be performed only after the basis for the phase has been demonstrated the *fundamentary regime* of the activity. (The demonstration itself may be part of the external activity.) I call an activity which satisfies the fundamentary regime *fundamentary*. Activities in the construction of rigorously grounded theories are particularly fundamentary. Aiming for maximal freedom and the elimination of all unnecessarily limiting rules, one must subordinate activity in the establishment of morality (i.e., systems of rules of conduct) to the fundamentary regime so that only grounded limitations enter into morality. The same is true in relation to legislative activity.

It must be noted that in a fundamentary activity a step not permitted by the rules of the activity cannot be fulfilled (since it is not authorized and therefore has no basis). Therefore the restrictive force of a fundamentary regime does not yield to the force of the despotic regime. In striving for freedom, therefore, it is impossible to impose a fundamentary regime on an activity when the freedom of that activity is under discussion although one must impose a fundamentary regime on the activity of establishing a morality for that activity.

In the fields of sociology and politics, the term 'despotism' is generally used in a sense which does not coincide with the one I attribute to this term in logic, although the two are connected. In these fields despotism means the presence of some 'despotic will' which subjugates the sphere of life being examined and prohibits everything that is not permitted and everything that it has the power to prohibit.

This will strives to make the realization of rights depend on the agreement of some one or other of the persons it has placed in power. In governing activities, the will of course prefers prohibitions to permissions in all cases for which it has not established authorizations, and in this way it follows the principle of despotism.

Under a despotic regime authorized acts can be performed without worry about prior indication of the authorization, but under a fundamental regime prior indication is required (although actions may be performed automatically). In this respect a fundamental regime is more rigorous than a despotic one.

According to the *restrictive clause* contained in definitions, a defined term may be used only in accordance with its definition. For example, the word 'bitch' may designate only a female dog. Imposition of a despotic or fundamental regime serves as a means of interpreting the restrictions expressed by the word "only." There are two different meanings of the word "only" which I designate as despotic and fundamental, respectively. Usage of the despotic meaning of "only" is connected with the elimination of the double negation. For example, if an object is designated by the word "bitch", then the lawfulness of this designation cannot be grounded. One can only assert that if this designation for the given object has not been authorized, then it has been either permitted and subsequently prohibited or not permitted and, consequently, prohibited by the despotic regime. But since such a designation could not have been implemented (in developing the definition under consideration) and since, by assumption, it has been implemented, then consequently the designation is not-not-authorized. This argument uses the law of the excluded middle, but if one wishes to derive authorization for the designation in question, the double negation must also be eliminated. Otherwise, when 'only' in the restrictive clause of the definition is understood in the despotic sense, one can merely assert that a bitch is not-not a female dog.

If 'only' is used in the fundamental sense, then, as soon as the word 'bitch' is used for designating an object, there must be a basis for this (connected with the definition of that word). Thus, the object must be a female dog. And so a bitch is a female dog if 'only' in the the restrictive clause of the definition is understood in the fundamental sense. Without going into further subtleties of the theory of definitions, I observe that activity in the use of terms introduced by definitions must be fundamental. Otherwise there will be no basis for using an assertion of the type, "A bitch is a female dog."

The juridical principle *nullum crimen sine loge* ("There is no crime without a law") corresponds to the principle of liberalism. In criminal law this principle is applied, but it only signifies permission for what has not been prohibited in that field of law. Without reservations or supplementary agreements, it would be risky to think that each person has the right to perform any act not forbidden. The problem is that the expression 'has the right' is used in jurisprudence in another connection as well. In civil law, for example, the principle that every right can be an object of legal protection must be defended. But it would be awkward to make some acts which are not prohibited objects of legal protection. The law does not prohibit anyone from becoming the victim of a crime, and at the same time the law cannot require a court to grant the suit of anyone who insists on his right to become a victim.

Perhaps this particular collision is cancelled by the distinction between active and passive voices, but there are other similar ones. The law does not prohibit resorting to necessary defense, but this right is not protected by the court in the sense that someone who has been prevented from making use of this right could demand from the court a reconstruction of the circumstances of the crime so that the right could be exercised. The rights allowing legal protection should be reviewed so that courts would have the means to offer this protection. To this end rights are brought into some system of "civil rights," but the right to every deed not prohibited is not included in the system. (At the same time there is no basis for extending legal protection only to rights stipulated by the system of civil law.)

It therefore follows that the presence of various legislative systems in states is called for by logical, not only social or historical, considerations. In addition, legislation must develop somehow harmoniously in deeper ways. Some of these represent morality, and thus moral systems of differing depth naturally arise.

I adopt the following principles in the theory of modalities:

If process E is described by method M, the rules of which contain the requirement that act A be performed in situation S, then for the continuation of E in S it is organically necessary that A be performed in S. (Principle of deontic-organic necessity, *pd-on*).

The same principle replacing organic necessity with epistemic necessity *(pd-en)*.

If in situation S of process E event C has occurred, then in that situation each event prior to C must have occurred. (Principle of ordinal necessity)

If in a possible situation S, A and B ... and K are possible, then A is possible in S (and B is possible in S, ..., and K is possible in S).

A and B ... and K are obligatory in a possible situation S if and only if, A is obligatory in S, B is obligatory in S, ..., K is obligatory in S. (Distributive principle)

In a possible situation S no violation of *pd-on* is possible. (Principle of negative evidence)*

For alethic modalities the following principles are adopted:

P1. For a possible situation S, if A is obligatory in S, then A takes place in S (or will take place in a later situation).

P2. For a possible situation S, if A takes place in S (or happens in S), then A is possible in S.

I will not dwell here on some more precise clarifications from the point of view of the logic of time which the formulations of these principles require.

For deontic modalities P1 and P2 can be violated, but this will violate the rules of the activity. P1 corresponds to the condition that in an activity which is going on any of its requirements is fulfilled. As soon as an activity is performed, this follows from *pd-en,* but this assertion is based on that which is necessary being considered obligatory (and on P1). If an agent is *loyal* to the rules of an activity, i.e., does only what is authorized ("only" in the fundamenty sense), then the condition corresponding to P2 is also fulfilled. For goal-oriented modalities the conditions corresponding to P1 and P2 can be considered two different characteristics of the "purposiveness" of the activity.

Theories of optative modalities are very important for the moral sciences if only because rules and ways are usually adopted for goals. They themselves serve as means toward these goals much as do the acts performed according to the rules or in harmony with the ways and the instruments or materials which aid in performing them.

The characteristic *evolution of goals* is that as soon as means M is involved in the attainment of goal T, goal T is replaced by a new goal TM: to attain T by means of M. So one desirous of acquiring thing E selects for this goal T means M, consisting of acquiring the needed sum

---

* The principles pd.-on. and pd.-en. have the versions "pd.-oc.," "pd.-ec." in which the necessity is replaced by compulsoriness (of the same group). These versions are stronger than pd.-on. and pd.-en. The principle of negative evidence for pd.-oc. and pd.-ec. is much less obvious than for pd.-on. and pd.-en.

of money. Often one falls into the *error of displacement of goals* consisting of calling M "necessary" or "obligatory" for T although, generally speaking, M is obligatory only for TM (on the strength of the definition of TM). A purchase, using money, for example, is clearly not a necessary condition for acquiring something. But of course the rules of external methods (in this example, rules of morality or legislation) may make M obligatory for T. In examining goal TM I will call M the *involving* means (to T). It is understood that for goal TM it may be necessary to involve an additional means M_1 and thus replace goal TM with goal TMM_1. This constitutes the evolution of goals.

A similar evolution is possible for desires. I see the distinction between goals and desires as being that for goals the agent seeks such means as he applies or desires to have the possibility of applying.

Generally speaking, a goal is not assumed to be attainable, and when this assumption is not made I call it an 'ideal'. This term generally applies only to the final goal of an activity which has subordinate goals set in separate stages or steps in the order of evolution described. Generally, applicability, i.e., the organic possibility of its application, is demanded of every means M in every situation S in which the means is applied for accepted goals. In particular, morality is commonly applied for the attainment of some goals, and therefore all its demands must be fulfillable. The same is true of legislation.

It is true that sometimes such means which have only the epistemic possibility of being applied are considered satisfactory. The choice of these means is connected with the risk that they will turn out to be inapplicable, and the less certain the aforementioned possibility, the greater the risk. When this certainty has been evaluated, all other conditions being equal, commonly the means associated with the least risk of this sort are preferred.

I call a situation S *hopeless* or a *dead-end* for goal T if it is impossible to attain T or (when T is considered an ideal) approach T in that situation. (The concept of 'approach' is defined more clearly in terms of preferences for some goals over others.)

I call the tactic of selecting a tactic for the attainment of a goal a *strategy*. I call a strategy *unswerving* if it allows a tactic, once selected, to be changed only in the following three cases: a) in the presence of the authorization of that change, for the current situation in which the tactic is being applied, i.e., in accordance with *pmf*; b) when the goal has already been attained; c) in a dead-end situation. In a hopeless situation changing the goal is also permitted, but only by rejecting the last means involved.

To *explain* some process (happening) means to indicate an unswerving strategy for discovering the method describing it. To *understand* a process means to find an explanation for it.

If only *honest* means are allowed for attaining a goal, then a morality, defining the concept of 'honesty' as steadfastly demanding it, is invovlved in this goal, and an application of dishonest means would immediately create a dead-end. Therefore such application could not be considered a suitable means for attaining the accepted goal. In the same way, if some judgment must be proved, the acceptance of any judgment on faith or the ungrounded acceptance of any rule creates a dead-end.

In the theory of goals I adopt the following principles:

To reach an unattained goal, sufficient means must be applied. (Inversion principle, *ip*)

To reach an unattained goal, every necessary means must be applied. (Supplement to *ip*)

Three principles of sufficiency:

1. To reach an unattained goal, it is sufficient to apply sufficient means for this.

2. Every event is sufficient for the occurrence of any of its consequences (and in this case the obligatory result of an event is called its 'consequence', but which results are to be considered obligatory must be specified for every event in each theory. For example, the consequences of writing a word are that the word is written as well as that the text completed by it).

3. Means M is sufficient for the attainability of goal T if there is an applicable way to reach T with the help of M (i.e., to reach TM).

These principles expand to cover all possible situations. (I do not adopt any principles for impossible situations if only because it is desirable to preserve the possibility of discussing the violation of any principle, but a situation can be impossible precisely because, according to its conditions, logical principles are violated.) In a more thorough examination the principles set forth here will need some clarification, but for the present they will assume a tautological grounding. For example, the i.p. includes in itself an agreement that realization of a goal without the application of sufficient means — let us say, in the case of a gift — is not considered the "attainment" of this goal.

With some reservations the following principles of transitiveness can be adopted: If B is necessary (sufficient) for T, and A is necessary (sufficient) for B, then A is necessary (sufficient) for T. (The reservations are not only related to the consideration of the situation. Clearly the sufficiency of A for B is not the same as the sufficiency of A for T, however simple it may be, having attained B, to reach T. But very often 'sufficiency for T' is understood to mean sufficiency for the attainability of T, and then the transitiveness of sufficiency takes place under very broad assumptions.)

Generally speaking, the occurrence of any event may be regarded as a goal. Goals are usually adopted on the basis of desires held or as a means for the attainment or the attainability of previously adopted goals. In this case the desires are selected according to the preferred tactic (taste) of the agent.

Roughly speaking, I divide *causes* of events (called *actions* or *effects* of these causes) into *eventual* and *necessitating*. For any situation S, A is called the *eventual (necessitating) cause* of B if, as soon as A has occurred, and only in that case, can (must) B occur in $S + \{A\}$. The elements of a situation S are called the *conditions* under which this cause *acts*. In Russian the presence of a causal connection is expressed by the words 'by this' *(potomu chto)* or 'by that' *(poetomu)*, etc., so that these modal characteristics most often remain unexpressed.

A *causal connection* is defined in an analogous way between phenomena distinct from individual events (processes or other enduring factors), between a phenomenon and an event, or between an event and a phenomenon. The principle, "The cause precedes its action," depends on the way precedence is determined and is often violated in cases where the cause is a prolonged phenomenon. (For example, although spring warmth is a cause of the growth of foliage, both phenomena develop simultaneously.)

I call the cause of an event's non-occurrence an *obstacle* (for the event). An obstacle is called *eventual* (*necessitating*) according to the modal characteristic of its cause. The more certain the possibility of the effect of this cause, the more serious the eventual obstacle (but, of course, an estimate of this certainty cannot be made in all cases).

The theory of modalities herein described bears little resemblance to contemporary logical-mathematical theories, and this is deliberately so, for it claims to lie at the base of these theories. At the same time it must form the basis of the logic of the moral sciences. Spinoza tried to attain rigorousness in ethics by means of axiomatic construction. I, on the contrary, try to attain rigor in the foundations of mathematics by means of a theory akin to the logical foundation of ethics. There are three very significant considerations by which I reject the axiomatic approach to this foundation: a) Judgments, or rules governing the derivation of some judgments from others, have been postulated; here one speaks of rules of an entirely different sort, and they are precisely the main subject under consideration; b) The dependence of the traditional axiomatic method on arithmetical assumptions about natural numbers, which have no foundation for being connected with the moral sciences; c) The contestability of any judgments selected as postulates for the theory. Instead of the axiomatic approach, I have adopted a definitional one, making acceptable judgments in accord with those which can be accepted on the strength of accepted definitions (or other rules of sign usage). This is achieved by means of the theoretical-modal principles considered above but they themselves are gounded in the rules of sign usage (as I have shown above for *pmf* and *ip*).

To be sure, the grounding of the theory of modalities under consideration has never been completed in detail. Moreover, this would not be an easy task, although it is considerably simplified when grounding the use of modalities in a finished logically well-thought-out text is all that is required. In this case there is only a limited goal, and there is hope of its attainment, although, until it has been attained, one must take into consideration the need for some modal specifications being included in the text or in its interpretation. This is how the matter now stands with the foundational studies of the Zermelo-Fraenkel system, where only the final text of the proof of the consistency of the system is subject to investigation (3). There is no such text in the moral sciences, but with any acceptably grounded fragment one could attempt to do the same. In this it must be necessary to clarify the fragment and change the rules set forth in it, but, unlike the foundations of mathematics, where doing so might violate a projected proof, in the moral sciences clarifying a fragment would certainly mean perfecting it.

There is, however, one serious difficulty in this way of grounding—the frequent dependence of accepted propositions on the elimination of the double negation. The identification of the necessity of A with the obligatoriness of A can be grounded only with the aid of this elimination. Without it, only not-not-A can be derived from the necessity of A. (For this reason I was not able to use 'necessary' instead of 'obligatory' in P1.) The non-impossibility of A can be examined as a form of the possibility of A and ground *pmf* for this possibility, but there is no basis for relating this possibility to the same group of modalities from which it was derived by means of the double negation. Although a proven judgment is by definition incontestable, only the identification of a contestable judgment with an unproven judgment clearly corresponds to the meaning of the word 'prove', so that what is incontestable, is simply not-unproven. Here lies a very deep problem which I will discuss in the extreme directions of my antitraditional program. The non-necessity of distinguishing judgment B from not-not-B for any accepted goal is an important means of overcoming these difficulties and gives the possibility, having rejected these distinctions, of replacing them with identifications.

Also the described principle of the logic of confidence, by the way, depends on the elimination of the double negation. And many problems in ethics and jurisprudence have to be related to it. Judges satisfy themselves with confidence in a witness's testimony based only on the fact that the law forbids a witness to lie on pain of punishment. But obviously, even assuming that the witness obeys this law and does not lie, only the fact that A is not false can be derived from his stating A; yet in general the judges accept A instead of not-not-A, and there is no basis for it.

But the fact that this commonly is not noticed shows such weakness of logic in the contemporary moral sciences that the level of rigor herein proposed as their grounding will undoubtedly appear a major achievement. In addition, this program for their grounding claims to subordinate the development of these sciences to some morality, but any morality, as has already been noted, must contain only fulfillable demands. In court cases the criterion of incontestability must also be understood in conformity with legal possibilities. In practice this particularly entails recognition of the impossibility of disputing testimony A on the sole basis that only not-not-A has been derived from obligatory testimony.

In establishing norms of courtroom procedure, besides the necessity of publicity (following, as was shown, from the principles of the logic of confidence), it will be necessary to pay attention to principles of the theory of relevance. Judges are granted the right to interrupt irrelevant speeches, but they must be denied confidence when there is a danger of their abusing this right. The concept of irrelevance depends on the tactics of attention: in any such tactic only that which cannot be connected with A (according to its rules) is called *irrelevant* for A, and the basic principle of the theory of relevance consists in its permitting only irrelevant matters to be disregarded. In this theory there are some general requirements for the tatics of attention. Attention to means toward goals often must be preferred to attention to obstacles in the creative stages, but in the control and executive stages this is impermissible, and obstacles must require undivided attention. In an honest dispute it is impossible to deprive either party of the possibility of proving the relevance of any subject to the theme of the dispute. In the courts the procedural rights of the parties must be considered clearly connected with disputes, and thus for a trial to be complete disputes about the law absolutely must be allowed in court. But an undue burden of proof cannot be laid on anyone (otherwise the rules could prove unfulfillable), and unfulfillable demands cannot be introduced into the procedure. Therefore some presumptions are inescapable, as is an acknowledgment of the lawfulness of using some generally accepted tactics, including tactics of attention. In some cases the irrelevant character of given subjects can be recognized as obvious. Further conclusions about irrelevancy can be made on the basis of the principle of alienness (applied, for example, in the following form: that which presupposes something irrelevant can be considered irrelevant). (This principle was introduced in [2,3]; for its theoretical-modal grounding see [3,pp.417-419]). However, presumptions may be considered incontestable only until such time as one participant in a dispute protests any one of them, citing a peculiarity of the case, the possibility of a general refutation, or any irregularity in their application. This right to contesting should be related to generally-accepted tactics just as it is to presumptions.

In particular, all cases of neglect not in accordance with generally accepted tactics should be clearly noted in a dispute since they can (and often do) play the same rôle as assumptions. For the same reasons all identifications made not in accordance with generally accepted tactics should also be noted in disputes.

The rules of the pure predicate calculus allow for a theoretical-modal grounding and therefore may be freely applied in disputes with only these reservations—that the grounding be connected with a definite interpretation of logical operators and that the laws of the excluded middle and the elimination of the double negation be inapplicable in disputed cases. But in cases when the applications of the latter laws involve constants or junction symbols, *iterated* applications of the rules of the predicate calculus must be accompanied by ultra-intuitionistic

A.S. Yessenin-Volpin

precautions [2,3].

By the term 'freedom$_1$' (*svoboda*)* mean the quality of acts of not being obstructed, i.e., impeded by obstacles; I call such acts *free$_1$*(*svobodny*). I call an activity *free$_1$* if in any of its situations every one of its acts is free$_1$, etc. I call an agent *free$_1$* if his activity is free$_1$. In this way the term 'freedom$_1$' signifies a quality of both an action and an agent.

In this case, in particular, 'obstacles' are understood as eventual obstacles.

The organic possibility of an act or an activity is compatible with the presence of an eventual obstacle which will not be realized. Therefore one may have the possibility of performing unfree$_1$ acts and carrying on an unfree$_1$ activity.

An activity encountering obstacles is not free$_1$, but if these obstacles are overcome, a wider activity, including overcoming these obstacles within it, may be free$_1$.

A free$_1$ act can be compelled. This often happens since a person compelling an act usually does not obstruct this act and may even eliminate obstacles.

I call the quality of an act's not being compelled its *freedom$_2$* (*vol'nost*) and the activity consisting only of free$_2$ (*vol'ny*) acts *free$_2$*,—in which case I ignore compulsions deriving from the requirements of the activity itself (i.e., describing its tactics). I call an agent free$_2$ if his activity is free$_2$ and if, in addition, he has not been compelled to choose it. I call this capacity in an agent his *freedom$_2$*.

A free$_2$ act may be unfree$_1$, and the same is true of an activity or agent.

Ordinary language uses these terms inconsistently, creating a powerful obstacle to their correct usage. Therefore, a term is needed designating the combination of freedom$_1$ and freedom$_2$; I will designate this combination by the Greek word *eleutheria*, and I will call acts, activities, and agents which are both free$_1$ and free$_2$ *eleutheric*.

Even this term is not felicitous in all respects. I call the absence of obstacles to the opposite act the *independence* of an act (understanding opposites as a pair of acts [A, not-A]—not-not-A-acts can usually be identified with A; in the contrary case the question becomes more complicated). I will call an act which possesses this property *independ-*

---

* Since there are no English terms which convey the contrast of *svobodny* and *vol'ny*, subscripts will be used: 'free$_1$' and 'free$_2$'

ent, an activity made up only of independent acts *independent*, and I will call the doer (agent) of an independent activity *independent* if the very choice of the activity is independent for him or if this activity is not selected by him and he did not have obstacles to prevent his selecting it. Acts compelled by the rules of an activity (including rules of external activities) are not considered as obstacles here. Independence is certainly a narrower quality than freedom$_2$ (i.e., an independent act must be free$_2$, etc.). Sometimes it is convenient to consider 'eleutheria' as the combination of freedom$_1$ and independence. I prefer to call this *eleutheria in the narrower sense*, keeping the previous meaning for *eleutheria*.

Morality can be established for the most varied purposes. It may be as hostile to the freedom$_1$ and freedom$_2$ of an activity as one could wish. But since one activity can obstruct another and even make it impossible, one must make a choice or prefer one to another.

That which is preferred is called *better* than what it is preferred to, which is called *worse*. For this very reason everyone always prefers the better and must do so on the strength of *pd-on*. This is a tautology. But different people are guided by different tactics of preference, or *tastes*, and even the same person at times follows different tastes, determining the choice of more particular tastes in various situations and in relation to various classes of objects being preferred). Cases where a person apparently chooses something worse are explained in this way. In such cases apparently there is a play on words, and the terms 'better' and 'worse' are applied in connection with a taste other than the one being employed. These terms are very often used to conform with a most widespread or common taste, and one must know how to correct this.

The very property of 'being better' is called *goodness*; the opposite property ('being worse') is called *evil*. These terms are also used to designate everything that is better or, correspondingly, everything that is worse.

If an object is preferable to some (similar) objects, and taste does not demand preferring another, that object is called *good*. On the other hand, when taste demands preferring some other (similar) object, such an object is called *bad*. To be sure, these words are sometimes used not entirely in this sense, but with intelligent usage their meaning may always be made more precise in an analogous way. Thus, 'goodness' is defined as the sum total of everything good.

Ethics is often defined as the "science of the good." But this word is used diffusely, and I prefer to define ethics as the *science of moralities*.

According to this view of the fundamental rôle of preference in any morality, this concept does not diverge greatly from the generally accepted one.

An activity can be a *field of action* or a concrete *act*. The difference consists in an operation's generally being made up of concrete actions thought of as being performed in some connection with each other, consecutively, etc., while a field of action consists of an abstract process formed by all possible acts of a given sort performed within the limits of conceivable activities. For example, a chess game is an activity (consisting of moves performed by turns according to definite rules), while the field of action of a chessplayer is the game of chess as such.

I will call the taste which prefers eleutheria to all other properties of an activity being considered *eleutheric* (in relation to that activity), and a morality chosen with eleutheric taste from among all other possibilities for a given occupation will also be called *eleutheric*. I will call the part of ethics studying eleutheric morality *eleutheric ethics*.

General motives prompt everyone to follow eleutheric taste in choosing a morality. Anyone engaged in an activity ordinarily is interested in avoiding obstacles, i.e., in attaining freedom$_1$ in this activity. One prefers to choose the activity itself according to personal taste simply because one is guided by personal preferences and wants to be free in this. Therefore everyone tries to avoid (extraneous) compulsions and strives toward freedom$_2$. This combination of striving for freedom$_1$ and freedom$_2$ defines *eleutheric* taste. But it may not appear so at all when a person is so deprived of freedom$_2$ that he does not choose his activity and is therefore compelled to be indifferent to it. In such cases even obstacles to the activity can not hinder the demonstration of personal tastes simply because these tastes are in no way associated with the activity. This is a *slave* relationship to an activity, but this phenomemon does not mean that the slave has no personal taste. Taste can be in evidence outside the limits of any activity. If the activity is compulsory, and the slave has been trained by it, indifference appears as one of the overriding features of this taste. Such a slave will strive to preserve his indifference, seeing goodness precisely in indifference, but even in so doing he will need freedom$_1$ and freedom$_2$, i.e., eleutheria.

In any morality, that which is permitted is considered a *right*, that which is required — a *duty* or *obligation* (although it would be more convenient to apply the latter term only in connection with legislation). Prohibitions limit rights, but I do not intend to say that they abolish them. Having adopted a morality, a person has, from the point of view of that morality, a right to that which is not authorized as long as it is permitted. The person must obey a prohibition, and then he cannot take advantage of a right. When in conflict with a prohibition, therefore, a right loses its actual force while retaining its theoretical meaning.

No morality exists until there are clear terms for the categories of rights and obligations.

An eleutheric ethic can be applied to the search for a morality for any type of field of action. (For some selected goals in the search for morality one must subordinate the field of action itself to an external activity.)

But in striving for eleutheria in a field of action, the search for morality (being also a form of activity) must be limited by considering any morality which contains ungrounded demands bad. The ethical law of sufficient reason consists in having every rule grounded. Since the initial taste in the search for morality is eleutheric, the search begins under the conditions of a liberal regime. Permission for any act in the field of action under consideration arises at once, but generally this creates a dead end, since some acts obstruct others (of the same or another person), violating their freedom$_1$. Limitations must be introduced. These primarily concern the freedom$_1$ of acts necessary for a field of action or for its goals, but also for the goals of activities being carried out in this field of action. The freedom$_1$ of some acts is attained at the expense of the freedom$_1$ of others, and the problems of the search for morality begin with the establishment of the necessary taste for freedom$_1$ or for eleutheric activity. Generally freedom$_1$ of needed acts is more important for the goals of a field of action than independence or even freedom$_2$ since the compelling of acts does not obstruct them. Nevertheless, eleutheria is necessary not only for the satisfaction of the agent, but also for the goals of the field of action since opposed acts, one of which is harmful for the goals (i.e., may serve as an obstacle to their attainment), can belong to one field of action, and in the absence of freedom$_2$ the harmful act may prove to be compulsory.

Eleutheria becomes a goal in relation to needed acts, but every demand upon the acts of the field of action may limit eleutheria. Therefore, the law of sufficient reason for these demands is adopted—every demand must be grounded. The proven judgment that a rule can or must be accepted for the attainment (or the attainability) of accepted goals can serve as grounding for a rule.

Thus every demand, including every prohibition, must be grounded. This means that during its construction a morality must be authorized for adoption only in the presence of bases. In constructing an eleutheric morality one must follow a fundamentary regime. But if accepting a demand again creates a dead end, exceptions, grounded on the presence of the dead end, are permitted. This is also in accordance with the fundamentary regime.

Demands are introduced only when there is a need for them, as under a liberal regime (the demand for an act is the prohibition of the opposed act). But in the presence of an accepted (grounded) demand only necessary exceptions must be made, as under a despotic regime.

Groundings for the acceptance of rules include two important groups — those sufficient for a goal and those necessary for it. (A third group consists of cases where new rules are adopted on the basis of previously accepted ones.) In the second case the rule absolutely must be accepted (on the strength of the supplement to *ip*), but one must also adopt those rules in the first group which are sufficient in their totality (on the strength of *ip*). As is true with every selection of sufficient means, the choice is far from simple. Sufficiency must be established for those means found; one way of doing this is by proving their sufficiency for an even more difficult, but better known, goal which, once reached, makes the attainment of the given goal obligatory. In this way sufficient — and more than sufficient — means are found for the given goal. Thus, although it is possible to choose among sufficient means (in the aggregate of such means), often one must choose those which are more than sufficient.

Demands included in an *eleutheric* morality may therefore prove to be redundant despite their being well grounded. This occurs when the grounding consists in proving the sufficiency of these demands for the avoidance of some evil. Therefore, according to eleutheric taste, even eleutheric morality must be constantly reviewed, whenever possible, with a view to easing its system of demands.

The search for a more rigorous grounding of morality or jurisprudence can be an important means to this end. For example, as a means of avoiding the appearance of especially dangerous murderers, a legislator is allowed to demand the severest penalty for murderers who are guilty of killing in a way that endangers many lives. Such a demand must be grounded in each individual case, but clearly this permitted demand is excessive in cases when it is applied to a murderer who has killed (only) one man and at the risk (only) of his own life. Formally such cases come under the rule of the aforementioned law; however, the law appears imperfect in view of the excessive nature of the demand permitted. Following the basis of the law, it would not be difficult to detect this excess and suggest that the legislator avoid it.

The distribution of the burden of proof in a civil or criminal trial is derived from the eleutheric principle of the necessity of grounding demands. The plaintiff must ground his demands by proving the presence of those circumstances which he advances as their basis. The burden of the proof of their lawfulness, that is, that in these circumstances, on the strength of the law, he is entitled to satisfaction of his demands, rests on him. The prosecutor, demanding punishment for the accused, must also ground his demand by proving the guilt of the accused. A court verdict assigning the accused a heavier punishment than that which the prosecutor was able to ground must be considered unjust.If the prosecutor refuses to prosecute,only a verdict of not guilty can be considered just (in as much as the court is considered only the evaluator of arguments presented to it, not the creator of new arguments).

I define *justice* (of a moral system, a law, an agent, etc.) as a following of the law of sufficient reason in adopting rules of conduct (in law or morality itself, as well as in any external character of conduct). This means grounding every demand and every exception to accepted demands, violating someone's eleutheria only on irreproachable grounds, and striving for a review of laws and morality according to eleutheric taste.

Principles of equality before the law, or equal rights, do not constitute the essence of justice, but are only an important feature of its development in our time.

Justice is greatly endangered by any allocation of power to some people over others or by granting of advantages which in fact give the possibility of exercising such power or in other ways suppress justice. Equalizing the rights and economic opportunities of people has been suggested as a means to fight this danger, but the principle of equal rights itself represents a not-fully-grounded demand (in view of its excessive nature) and in practice it is observed only with reservations. Moreover, in fact legislation necessarily either violates or emasculates this principle each time a question involves the rights of people occupying opposing positions in a system involving the exercise of power. What is there to say about the equal rights of a commander-in-chief and a common soldier? They are equal before the law, but only in the sense that each one, being a commander-in-chief, has the same rights and responsibilities, and likewise for a soldier. (This way of conceiving equal rights constitutes what I have called the "emasculation" of that principle.)

But despite the imperfection of the principle of equal rights, contemporary seekers after justice as a rule are bound by this principle. The danger just mentioned continues to exist and reappears at each departure from the principle. It of course follows that one must achieve precision in formulating reservations to this principle, but searches for a better principle, though considered, have not become actual thus far.

The demand that everyone tell only the truth—or, even more, the whole truth—in all cases could be considered just. The danger of lying and silence being used as means to cover up an injustice would serve as a basis for this. But upon reflection anyone would recognize such a demand as unfullfillable since ordinarily people do not know the whole truth and are constantly unable to escape various errors and inaccuracies. Besides, in some cases frankness turns into impermissible betrayal. Therefore this demand must be limited to the prohibition of an intentional *lie* (i.e., expression of opinions which are incompatible with the opinions of the speaker), and silence must be permitted in all cases other than those where it is incompatible with the obligations of a given person. Furthermore, it would be absurd to accuse a mathematician who publicly adopts a false assumption in order to refute it or an actor who calls himself "Hamlet" on stage, although that is not his name, of lying. In addition to intention, indicators of an intentional lie must include the lie's being committed without warning and without the consent of the interlocutor as well as its being directed generally toward a person who has the right to hear the speaker. Anyone who has given his audience or readers warnings and reservations clearing his words of any false character must be free of reproach for lying even if, through no fault of his own, the warnings and reservations are not heeded by his addressees. One who has himself, by morally inadmissible means, compelled another to lie has no right to rebuke the other for lying. With these qualifications, an ethical prohibition of lying can be considered just. In any case, it is more just than the previously considered demand to always speak only the truth, and it does not infringe upon the freedom to express any opinion, which is the chief value of freedom of speech.

However, one can agree that, despite the reservations stipulated, this demand never consciously to lie (which now assumes "without warning and without the consent of the addressee") remains excessive, since a lie does not always infringe on someone else's rights or present a danger, and this prohibition limits freedom of speech all the same. Thus it would be even more just to demand the prohibition of lying as a means of concealing acts which are at variance with the demands of a recognized morality, as well as the full prohibition of lying in the course of procedures directed toward deciding disputed questions of morality or law. Without pursuing this theme further now, I merely call attention to the way that successively considering bases in this example led to a restriction of the demand and an increase in its justice.

Common morality consists of rules either wholly unwritten or adopted by a religion or other ideology and constantly undergoing arbitrary interpretation and factual distortion. The basic task of legislation is precisely to fixate formulations which are obligatory for all.

The establishment of morality, however, is not solely a social task. Every individual in striving to maintain his own character is compelled to invent rules of conduct in various circumstances to attain this goal. Such rules make up a system of *personal* morality, and an individual can also follow a variety of personal moralities in various fields of action or under essentially different circumstances. The individual himself decides these questions and, in the work of ethics, enters into the study of useful recommendations. An individual may subordinate his conduct to the rules of a written morality which he establishes.

Legislation may contain *norms* appearing not simply as rules, but as rules provided with an indication of their goals and supplementary rules *implementing* a definite way of following the basic rules, i.e., preventing their violation. Supplementary rules can also anticipate "compensatory conduct" in case of a violation of a basic rule, i.e., in such a case demand the adoption of some other (usually defined) system of rules in pursuit of the same goal.

From the viewpoint of eleutheric taste, the general goal of the adoption of legislation must be to include all rules and other norms necessary for eleutheria while allowing each individual to ignore any rules (norms) not included or choose among them according to his own taste. The demand for publicizing legislation derives from this, since without publicity the attainment of this goal would be hindered.

One must remember that morality and legislation always set themselves other goals besides eleutheria. Often they do not fully value the meaning of freedom$_2$ and try to provide freedom$_1$ only for those fields of action or agents which the lawmaker considers useful$_1$ for

---

*I make the concepts of use and harm explicit in the following way: to *aid* in attaining a goal means to give or create means for its attainment or to remove obstacles (to give or create means for their removal). Everything that

satisfying his desires. This is a violation of eleutheric taste.

Many norms included in legislation and morality are essential not for one field of action or another, but for life as a whole, and in such cases life must be regarded as the broadest field of action of a person, embracing all other fields of action.

When eleutheric taste is violated, the freedom$_2$ and especially the freedom$_1$ of the affected fields of action are not so much ignored as considered subordinate to the other, more important goals for these field of action. Violations of freedom$_2$ can really be useful for a field of action although they may harm the people participating in them (i.e., their desire to follow their own character). A lawmaker (or one who establishes such a morality) prefers the interests* of the field of action affected to the interests of the people participating in them.

It is evident that eleutheric taste must be expressed in quests for means of its proper implementation, means for preventing any violation of it. But this does not necessarily mean that eleutheria is acknowledged as the supreme goal. Often people rationally consider it only a means for attaining other, higher goals such as striving for truth, love, success, etc. However strong eleutheric taste may be, it must, according to its own rules, allow these strivings to develop. Their very development is needed for eleutheria.

---

can aid in its attainment is *useful* for a goal; everything that can create obstacles to its attainment is *harmful*. *Usefulness* and *harmfulness* are the capacity to "be useful" and "be harmful," or the aggregate of what is useful or harmful (sometimes also called 'good' or 'evil'). According to the theory of relevancy, in each case only relevant use or harm is considered. When applied to desires, use and harm are understood as use or harm in the attainment of the goal of satisfying the desire.

*That which is connected with accepted goals and desires is *interest(ing)*. Something interesting *offers interest*. But when one speaks of preserving or preferring interests, only interests useful for these goals and desires are considered.

But people are often mistaken about the meaning of their own strivings. The concept of 'truth' is connected with the acceptance of faith, which immediately creates a danger of error and inclines toward satisfaction with an incomplete knowledge of truth. People strive toward a universal love, forgetting that before all else love is a preference for the beloved creature above all others and therefore cannot be universalized. They tell each other that various ideas are good, hardly caring about the bases for using this term. Contradictory stituations are created in which incompatible goals and desires are recognized as good; obstacles necessarily arise and freedom$_1$ for many important and highly valued goals vanishes. In such circumstances a person who knows how to adapt well to common prejudices and manage their play often gets power over the others. This is a case of "ideological power" (understanding 'ideology' as a system of prejudices significant for morality*). Without necessarily going further in its choice of goals than striving to subordinate such power and any conflict ensuing to the norms of eleutheric morality, eleutheric taste must either escape from or fight with ideological power. Striving for just legislation regulating the course of the conflict and manifestations of power, not striving to end the clash between ideas, expresses eleutheric taste.

* * *

Various ideological moralities have as goals the attainment of mutual aid among people. Eleutheric ethics must strive not so much to suppress these goals as to attain recognition of the necessity of first teaching people not to disturb each other. Every eleutheric morality must come from this principle; it expresses striving for eleutheria. But the search for an eleutheric morality is complicated, and, until it is found and established, eleutheria can be the highest goal of the seekers. During this search one must temporarily be guided by one or another

---

*The word 'ideology' literally can signify any system of views, including one based on logic. But in fact it has long ago lost that meaning since logically-grounded views, in the course of becoming known, cease to be objects of sharp social conflict—people agree with them. In "ideological warfare" those views which are based only on the tastes of the combatants, before any examination, are precisely the ones which acquire sharp significance. For this reason I consider ideology as a system of prejudices.

preliminary system of eleutheric morality including appropriate rules of mutual aid and other aspirations destined to be developed once a satisfactory eleutheric morality and legislation beome general, but these preliminary systems are imperfect. Regarded as a means for eleutheria, they must be developed and perfected in accordance with an evolution of goals, steadfastly striving for eleutheria. The very concept of eleutheria must be made concrete in the course of the examination of changing fields of action.

Eleutheria is a means for a wide range of goals, and its worth consists in its securing many norms which are also such means. These means are necessary for many goals; although taken separately they do not suffice to attain these goals, in totality they can prove sufficient. When a choice among several such means becomes necessary, this must follow from not every means' being sufficient. Therefore if a choice must be made to adopt one of two (or more) means, preference must be given to the one which will make it possible to adopt the other in time. This is the substance of the *principle of rational choice* among means (applied, it is understood, only when one of the means considered can provide this possibility).

This principle is frequently violated for the sake of another—the *principle of urgency*—which demands that the means of obvious usefulness in a given situation be selected. This preference for the interests of the present over those of the future can only be rationally grounded when violation of the interests of the present threatens the very continuation of a field of action. Under eleutheric morality and legislation, the continuation of protected fields of action is always worthy of concern, and, with the exception of the aforementioned cases (at least when the threat is serious), these fields of action must conform to the principle of rational choice.

In the field of recognition of rights, this principle appears in preferences for authorizing the use of those rights which can aid in protecting other rights. Even if only rights which are essential in daily life have an urgent significance for people, still preference must be given to recognizing those civil and political rights which are necessary for the protection of these urgent ones. But in circumstances where the very continuation of daily life is faced with a general threat (war, strife, or natural disasters), restrictions of these rights are permissible in order to meet the threat. Important for any right is this idea of extreme conditions under which recognized rights can be restricted, though only in the way described.

In accordance with the principle of rational choice, eleutheric ethics must uphold the following hierarchy of rights:

The right to the defence of every right must be recognized as the supreme right. In relation to unacknowledged rights, this recognition must include within it recognition of the right to fight for the assertion of unrecognized rights in morality and laws. In this case a right constrained by a recognized prohibition should be considered 'unrecognized'. In relation to recognized rights, this supreme right must include recognition of the right to fight against any threat to violate recognized rights and to fight for their restoration wherever they have been violated. This right itself must be recognized unconditionally although its usage must be limited by the demands of eleutheric morality (a stipulation applied to all rights; I shall not repeat it in what follows).

Following the principle of rational choice, one must recognize the next most important right: the *right to freedom$_1$ and freedom$_2$ of thought*. This right assumes recognition of the *right to life and health*.

The rights just mentioned must be discussed with a clear delimitation of rights and obligations. Life, a most important gift, must be not only free[1] but also free[2], meaning that life should not be turned into an obligation for the living. In order to implement the distinction between a right and an obligation in this sphere, I propose to accompany recognition of the right to life with recognition of the *right to suicide*. Any general eleutheric morality must contain such a right; the personal morality of each individual can prescribe it as a duty to continue fighting for live as long as life has value for him, but eleutheric ethics, not the morality accepted by society, must prescribe the right to suicide.

Even participation in the life of society, choosing one or another morality for oneself, must be free$_2$. Recognition of this principle must be considered a necessary corrective for any imperfections of morality or legislation. Therefore the *right to leave a society* must be recognized for each person, limited only by the demand that the fulfillment of obligations stemming from just demands of the morality and laws of that society be guaranteed.

Furthermore, the *right to associate with other people* for the goal of lawfully attaining one's goals or to guarantee the freedom$_1$ to realize one's rights must be recognized for everyone. This right can be limited only by just demands directed at preventing the formation or development of associations which endanger eleutheria.

Additional civil and political rights (to freedom$_1$ of travel, expression of opinion, meetings and assemblies, choice of one's field of activity and one's role in it, fair trial, etc.) must be counted by legislation among the inalienable rights subject only to those restrictions without which the lawgiver cannot guarantee their observance. Various social and cultural rights (to education, work, leisure, compensation, and material security) become necessary as society develops and life becomes more complicated for the participant. The lawgiver must strive, insofar as he is able, to recognize and guarantee these rights (without limiting the freedom$_1$ of those people who refuse to carry the associated burden to leave the society). The development of science increases the possibility of public health care, and the lawgiver, in the course of protecting each one's right to life and health, must strive to recognize everyone's equal right to make use of this aspect of progress.

In the field of criminal legislation, the lawgiver must confine himself to prohibiting those acts which obstruct recognition or realization of the rights of individuals or lawful associations. These prohibitions must be considered limitations of civil and political rights. They should be expressed by norms which present serious obstacles to the commission of these acts (called 'crimes') by creating inescapable threats for those guilty of them. In choosing these threats the lawgiver must go no further than is necessary for the loss of the desire to commit a crime (or to cooperate in its commission) on the part of a potential criminal.

Contemporary criminal law establishes punishments which do not at all correspond to this goal and are connected with crude, often unmotivated prohibitions of freedom$_1$ of movement. Logic can only seek explanations, not justifications, for these measures. It is futile to try to logically ground some crime's having to be punished by, say, five years' deprivation of freedom instead of four or six since the very measure of deprivation of freedom in the overwhelming majority of cases represents an injustice. The contemporary way of enforcing this represents an additional grave injustice, forcing one to see the makers and executors of laws as the enemies of all eleutheria. But logical connections could be established between the gravity of various crimes and, correspondingly, the severity of punishments. Although rape is a very grave crime, the law which punishes it more severely than premeditated murder is clearly unjust. Logic can assist in removing such imperfections in legislation.

Criminal laws act to restrict civil rights, but the legislation must be public particularly so that each one can know the limits of his rights.

Therefore criminal laws must be formulated with sufficient clarity to guard the rights they defend and at the same time not subject other rights to unnecessary constraints. As I have already stated, the necessity of legislative activity constantly gives rise to some excess in prohibitions, but one must strive to keep it to a minimum. Every inexactness in criminal law must be interpreted by the courts to the advantage of the accused (this is one manifestation of the presumption of innocence), and laws must be drawn up taking this circumstance into account.

Legislators of criminal law should recognize that in individual cases the prohibitions they establish may in practice prove harmful to these goals or to more important laws, and therefore they should provide that violating the prohibitions in such cases does not represent a crime (principles of extreme necessity and necessary defense).

The legislator is compelled to forbid not only that which he must consider a crime, but also that which creates the danger of crimes. This is clear since the chief goal of criminal legislation is precisely the prevention of crime. Thus in many cases an attempt at a crime must be considered a crime. But it would be dangerous to go too far in this direction since in practice an attempt may be represented by harmless acts. For example, acquiring a fatal poison for the purpose of murdering another person can be considered fully grounded as a dangerous attempted crime. But punishing someone for an attempt to commit this crime is equivalent to punishing for an attempt at an attempt. A visit to a pharmacy (even without a prescription, in order to make inquiries) can be accounted an attempt to acquire poison. And so approaching a pharmacy can be considered an attempt at such an attempt. Of course one must stop at some point, and in defining this stopping point it must be remembered that even keeping poison in the house is not always connected with serious danger to anyone's life.

Therefore cases of punishable attempts at or preparations for a crime must be defined precisely by law without causing unnecessary restrictions. In this connection it is useful to establish that attempts and preparations are punishable only if a punishment is included under criminal law. This must also be established for various forms of indirect participation in crimes (complicity, instigation, non-reporting, concealment, etc.)

One danger in establishing oppressive criminal prohibitions is the indifference of the public, and often the legislators as well, to the 'authorization' of acts which are in practice either impossible or rarely necessary. People rarely emigrate abroad and may therefore lose a vital interest in this right. As a result, completely contrary to the principle of rational choice, prohibitions of and insuperable restrictions on this right can be introduced and accepted almost painlessly. No one is capable of jumping over his own house and the public could accept a prohibition against doing this lightly. But if the law forbids any attempt at a crime, and the judicial process does not protect the rights of the accused sufficiently, anyone who for some reason ran toward his house and was suspected by a policeman of running with the intention of trying to commit this crime would feel the weight of that prohibition on himself.

The specification of criminal prohibitions serves as a guarantee of the observance of civil rights. But such prohibitions do not have to exhaust the number of guarantees. For all isolated cases of violation of civil law, a procedure of judicial satisfaction of civil suits authorizing restoration or compensation must be anticipated by legislative process. Such violations must either not be considered crimes at all or, when they endanger the further functioning of the legal order, be subject only to a fine or milder measures (besides the satisfaction of the civil suit).

The principle of time-limit [statute of limitation] must be recognized in criminal law. The bases for recognizing this principle are the recognition that people change with time and that it is clearly unjust to punish someone who was not guilty of something. The factor of time can play a more important rôle in identifying or distinguishing two personalities than common origins or external resemblance and physical identity. This consideration is completely independent of the sort of crime to which it is applied, but simple formulations can only lead to very crude and unsatisfactory criteria. Of course, after a lapse of twenty years it is awkward to punish a person for any crime whatsoever (the term of concealing a crime from judicial action can be included in the period of the crime), but criteria allowing the grounding of a much earlier application of time-limit must be sought. However, a point of view demanding (always or in serious cases) that for a certain time limit the criminal not be freed from responsibility, but the question of the permissibility of the application of time-limit be raised, is possible. In this case doubts must be resolved in favor of the criminal since they are doubts as to the correctness of his identifiation.

I have already spoken of the importance of judicial procedures. Since they are guarantees of the correct definition of the administration of justice, they themselves must be carefully described by rules of trial procedure drawn up with the demands of justice (including assignment of the burden of proof and, in connection with this, the presumption of innocence) and the principles of the prototheories, in particular the theory of relevancy, taken into account. On the strength of the principles of the logic of confidence, publicity, as I have said before, must be considered a necessary requirement for judicial procedure. Since violations of publicity create the possibility of further violations of procedure being easily committed, publicity appears as a most important and necessary means of fighting for the implementation of all the remaining rules of the administration of justice, and, on the strength of the principle of rational choice, it must be considered the most important of the procedural principles.

In spite of this, legislation concerning trial procedure is forced to introduce some restrictions on the publicity of courts. But, like every violation of necessary conditions, these restrictions must *in all cases* be compensated for by special guarantees that attain the goals of these same conditions. Thus *any* violation of the publicity of the courts anticipated by the law must be compensated for by additional guarantees *again based on publicity*. (I explained how this might be done in my work on publicity published in issue No. 7 of the journal *Obshchestvennye problemy* [*Social Problems*], 1970) Publicity is only the most important condition, but it is far from the only one necessary for justice in judicial procedure. The violation of any such condition, or of any important though not necessary condition, must be considered sufficient indication that a court has not judged, and this must render any decision adopted null and void.

The publicity of legislation must be expressed not only in generally available publications, but also in codifications which give everyone the opportunity to verify that some law exists and, in case it is missing, to refer to this fact as a manifestation of the will of the lawmaker. Codes are sufficient for this purpose; collections put together by jurists are not always sufficient. The constitution of a state must contain a list of the codes governing the extent of the rights and obligations of each person.

To attain the necessary precision each code must contain a section presenting the meanings of all words used in it which are little known to the public or are ambiguous in ordinary language as well as rules for interpreting the laws contained in it.

Striving for brevity is useful in legislation since brevity aids in locating a law and contributes to the portability of the code, but clarity and completeness of formulations must be considered more important qualities of legislation. Therefore one must not fear lest the code become a heavy folio volume, although for everyday purposes in such cases abbreviated editions or partial codes and fragments of whole ones, may be necessary.

In the field of morality I agree that falsehood [lying] is the most dangerous of all human vices. Guided by the principle of rational choice, I recommend that the fight against this vice be carried on everywhere as a means of overcoming the remaining vices. Let each person dare to sin only to the degree that does not require him to lie in order to conceal or justify his acts. If this principle prevailed, all other dangerous vices would be rendered sufficiently harmless, but as long as lying is authorized, all other means of fighting for moral perfection will hardly be effective. Therefore I consider the problem of the fight against lying the basic one in the task of implementing any morality. But the means of fighting lying must be just. Each individual must be protected by legislation from false accusations of lying, and the lying itself must be prosecuted by means more just than criminal penalties. The natural punishment for a liar is to be exposed and (for a reasonable period) denied confidence. This is the means that eleutheric ethics must recommend to any eleutheric morality. Every supporter of eleutheric morality must steadfastly recommend that a prohibition against ethically impermissible lying be included in one's personal morality. The more precise definition of this prohibition must be reserved for theories, but in practice I think every honest man can link himself to this without serious discomfort and not consider it extremely restrictive.

The development of the custom of employing any rule precisely as it is adopted is another important means recommended by eleutheric ethics for implementing any just eleutheric morality. The words entering into the formulation of rules must be understood only in their literal meaning, and the same must be true of accusations, reproofs, and demands presented to other people. When I say this I have in mind the necessity of overcoming the habit of associative thinking in all these cases, however valuable the capacity to associate ideas may be in various fields of creative work. Justice consists in using rules and constraints in accordance with a fundamental regime, i.e., being always well-grounded; association has value only in the creative stages of an activity which are subject to a liberal regime. (Of course this does not relate to the kinds of association used for understanding the exact sense of various texts, including rules.)

In particular, one must distinguish between accusations and reproaches, using the first only in connection with crimes or demands for punishment. When a bad act does not give bases for accusations, one must be limited to a reproach, that is, an indication of the bad aspects of an act which does not entail any consequences beyond those foreseen by the rules of the field of action and, perhaps, the breaking off of personal relations.

Many imprecisions inherent in everyday speech and dangerous to justice must also be fought. In Russian, for example, the phrases "does not have to" [*ne dolzhen*], "it is not necessary" [*ne nado*], and "I do not want" [*ne khochu*], despite generally recognized rules of grammar, are used in the sense of "has to not" [*dolzhen ne*], "it is necessary not" [*nado ne*], and "I want not" [*khochu ne*] more often than in their literal senses. (Analogously for the common colloquialism "it is not commanded" [*ne veleno*].) This removes the possibility of using negations and modal words as signs in logical deductions and inclines one toward associative or figurative thinking where logical precision is needed. Another example of a dangerous imprecision in language is the widely disseminated disregard for the word 'only' which is often omitted where it ought to stand—even in laws and other judicial documents. One could adduce a good many similar examples. In necessary cases, of course, precision can always be restored, but one must insist that this always be observed in judicial documents.

Moralities and legislation always face the problem of who is affected. While discussing the principle of equal rights, I spoke of the dangers of discrimination. Because of these dangers it is preferable that the effect of every eleutheric morality affect all who agree to that and are not bound by the principles of another, incompatible morality. This means recognizing the rights of foreigners in particular. Of course this principle must be applied with limitations directed against the dangers of tolerating too many subjects alien to it who would want to subordinate the principle to their own ends or destroy it. Legislation must recognize the freedom of association within a society with the same limitations.

It is not necessary to seek a logical grounding for justifying existing social and government institutions since these institutions will certainly change and, in many cases, should on principle, in justice give way to better ones little resembling existing institutions. The state itself has value only as a legal institution guarding the rights, interests and lives of its citizens and other residents of its territories. The state claims powers not connected with its rôle and by these unjust claims it naturally turns honest citizens into enemies. In developing eleutheric theories, one must separate the state as a legal institution from all other functions of those organisms which have been called "states" in history and politics. As a rule, when connected with these other functions, demands on individuals and associations are unjust.

The territorial principle of extension of governmental power is defined not by the demands of justice, but by historical circumstances. A free distribution of societies following various just systems of morality and legislation in their inner lives and intermingling in cities is fully imaginable. Isolated excesses—conflicts and crimes — would be considered by local powers independent of these societies and capable of applying the norms of any law depending on the affiliations of the parties or on agreements between the societies. But this assumes a level of ethical development higher than that reached by any human society up to the present time.

I have set forth everything on the theme of the logic of the moral sciences that I consider possible and sufficient for the present article. Of course this is not a logical investigation in all respects since I have deliberately permitted some freedom of style. A logical investigation of this theme would have taken up a great deal more space and would have appeared to be a clarification presented in the spirit of extreme pedantry. The basic idea of this work must be that the rules of ethics and jurisprudene can admit a far stricter grounding than that heretofore accepted. This basis, relying only on generally accepted tactics of sign usage which are unavoidable in any dispute, is the focus of the discussion.

Of course, not everything in the moralities and laws adopted in one place or another allows for such grounding, but this is so only while the goals of a legislator remain unexposed and the circumstances in which he creates his rules unclear. With the clarification of either point, it will always be revealed that a rule is accepted in laws as a means for attaining the goals of the society or the legislator in given conditions, and that this happens in accordance with optative logic. This is so if mistakes, as unavoidable in legislative activity as in any other, and manifestations of obvious, sometimes absurd, petty tyrannical power are avoided.

Much that is valuable in my eyes in this conception of ethics and jurisprudence seems trivial and generally known. Just rules and systems should become trivial and generally known; this is the task of thinkers studying them. This prepares the way for their acceptance and observation in places where these goals have not yet been attained. But

I want another goal to be attained as well. The development of morality and law have been most often explained by historical, sociological, psychological, biological, geographical and political factors. The logical factor has hardly been mentioned. This promoted the spread of the conviction that the victory of just norms is always connected with some sort of historical conditions. Now let logic in its development show everyone what freedom$_1$, freedom$_2$ and justice are, and may the very spread of the ideal of eleutheric ethics constitute an important historical and social factor preparing for the victory of those ideas.

December, 1970
Copyright© Alexander S. Yessenin-Volpin
Moscow, 1971, Boston, 1988.

List of abbreviations

(the numbers indicate the pages on which the designated principles are formulated)

pmf	principle of modal fulfillment	page 118
pd.–on	principle of deontic–organic necessity	126
pd.–en	principle of deontic–epistemic necessity	126
pd.–oc	principle of organic compulsoriness	127
pd.–ec	principle of epistemic compulsoriness	127
i.p.	inversion principle	129

Bibliography

1. A.S. Yessenin–Volpin., "O teorii modal'nost'ej", in the volume *Logika i metodologiia nauki*, Moscow: Nauka, 1967, pp. 56–67.
2. A.S. Yessenin–Volpin, "The Ultra–intuitionistic Criticism and the Anti-traditional Program for the Foundations of Mathematics," in *Intuitionism and Proof Theory, Proceedings of the Conference at Buffalo, 1968*, North–Holland, 1970, pp. 3–45.
3. A.S. Yessenin–Volpin, "Ob ul'tra-intuitsionistskom obosnovanii sistemy Tsermelo-Frenkelia", p. I, Moscow, VINITI; deposited manuscript no. 1202–69, 450, Yessenin–Volpin, December 1970.

Bibliography

1. A.S. Yessenin–Volpin., "On the Theory of Modalities," in the volume *Logic and Methodology of Science*, Moscow: Nauka, 1967, pp. 56–67.
2. A.S. Yessenin–Volpin, "The Ultra–intuitionistic Criticism and the Anti-traditional Program for the Foundations of Mathematics," in *Instiutionism and Proof Theory, Proceedings of the Conference at Buffalo, 1968*, North–Holland, 1970, pp. 3–45.
3. pt. I, Moscow: VINTI, 1969: deposited manuscript no. 1202–69, 450 pgs. (In subsequent parts, the theory of modalities mentioned therein and in the present work will be further developed).

Philosophy of Concept Art

An interview with Henry Flynt by Christer Hennix

Dec. 6, 1987

FLYNT: I'm going to give a summary of how I originated Concept Art in order to bring it up to the point where it's understandable why I speak of you (Christer Hennix) as my only successor in the genre. Summarizing briefly, I see two things coming together. One of them was my involvement with the modern music community of the time—Stockhausen, Cage, LaMonte Young—and the other aspect was that I had been a mathematics major at Harvard and already knew that I thought of myself primarily as a philosopher—that my intention had been when I was very young, when I didn't understand the situation that I was in—my intention had been to become a philosopher with nevertheless a specialization in mathematics. Of course, many people actually did that.

So, having said that, one of the things that I began to notice about the modern music of that time was this extremely strong pseudo-intellectual dimension in Stockhausen—Stockhausen's theoretical journal *die Reihe*—the impression that they were doing science actually—for example Stockhausen had a long essay on how the duration of the notes had to correspond to the twelve pitches of the chromatic scale...

HENNIX: "...how time passes..." [*die Reihe* 3]

FLYNT: Yes, and what is more, the other rhythms had to correspond to the overtone structure above those frequencies as fundamentals.

HENNIX: Yes, I'm quite familiar with that.

FLYNT: Yes, I would expect you would be. I remember Bo Nilson—you will like this—in 1958 at the same time I saw Stockhausen's score—he went even one step further than Stockhausen because he used fractional amplitude specifications—so this is even more than Stockhausen, and so forth and so on.

Cage took a considerable step further in the sense that in Cage this kind of play with structure is carried to the point where there is an extreme dissociation between what the composer sees and what the performer sees in terms of the structure of the piece and what the audience knows. They are completely divorced from one another. Cage would compose a piece on a graph in which the time that a note begins is on one axis and the length of the note is on another axis. What he would do was to superimpose that on some picture like from a star catalogue—

HENNIX: *Atlas Eclipticalis—*

FLYNT: Yeah, well, that's the particular piece. I'm making up a composite of his compositional techniques but the result is that when you break up a sequential event in that way, it's not like a pitch-time graph where there's an intuitive recognition of the way the process unfolds. He would have one structure for beginnings and another structure for durations. Well at any rate, already in Cage's music there was a kind of ritual aspect to performing classical music. I mean in Cage's piece, which is actually all silence—the only thing the pianist does is open and close the lid of the piano or something like that.

Then LaMonte Young comes along. His word pieces were the first that I ever saw, composed in mid-1960. I saw them in December 1960.*
It was a very different kind of structural game. It was no longer like twelve-tone organization and so forth but rather it was like playing with paradoxes—it was nearer to making a paradox than making some kind of complicated network.

And I felt that matters had reached the point where there was some kind of inauthenticity here because the point of the work of art had become some kind of structural or conceptual play, and yet it was being realized under the guise of music so that the audience had no chance of really seeing what was supposed to be the point of the

---

*Other composers have earlier dates, but for me, Young crystallized the genre. [H.F., note added]

piece—the audience was actually prevented from seeing. Certainly Cage's methods had exactly that effect. The audience receives an experience which simply sounds like chaos but in fact what they are hearing is not chaos but a hidden structure which is so hidden that it cannot be reconstructed from the performed sound. It's so hidden that it can't be reconstructed but nevertheless Cage knows what it is. So I felt that the confusion between whether they were doing music or whether they were doing something else had reached a point where I found that disturbing or unacceptable.

At the same time at that period there was a great fascination in sort of taking the Stockhausen attitude and looking back at the history of music from that point of view. Stockhausen's analysis in *die Reihe* 2 of Webern's String Quartet [Op. 28] tried to show that Webern was composing total serial music and not just twelve tone music. That was the attitude, they were rewriting the history of music, trying to show that all previous important figures were essentially preoccupied with structure, that they had been complete structuralists.

HENNIX: Really? I thought it was only Webern that was given that treatment.

FLYNT: Well, they were digging up all these composers from the Middle Ages, the isorhythmic motet and everything like that—they were sort of dredging that up because that was the previous period—the medieval scores in the form of a circle and the use of insertion syncopation,* it appears with the red notes in a medieval score and then it reappears in Stockhausen's *Klavierstück* XI. They were just jumping, they were dismissing what we would call the baroque, classical and romantic periods periods as completely worthless. In other words, the last music before Stockhausen was in the 14th century, this is the way the history of music was being rewritten. And LaMonte was getting into Leonin and Perotin and all that kind of stuff. Well, anyway, that's quite an excursion.

At any rate there is in music, there is this preoccupation with—it may be a kind of quasi-Pythagoreanism, I don't know ...

HENNIX: The way I looked at it was that they saw in Webern, first of all the harmony was going away. And they saw in Webern a way of

---

*My term for the rhythmic feature common to Magister Zacharias' *Sumite Karissimi* and *Klavierstück* XI. See Willi Apel, *The Notation of Polyphonic Music* (4th ed.), p. 432 for *Sumite Karissimi*. [H.F., note added]

determining the note more and more precisely, in terms of all of its parameters, pitch, duration, timbre and all that. What was left was that timbre was not serialized yet. And that, as far I see it, was what the Darmstadt school did—they added—

FLYNT: Stockhausen's *Kontra-Punkte*—

HENNIX: Yeah. And they all considered Webern the god of the new music—

FLYNT: Yes—

HENNIX: —and also a little bit Messiaen—

FLYNT: Yes.

HENNIX: It was Webern and Messaien that determined the entire fifties in Darmstadt. In other words, they were saying that Cage was no good. He was just looking in *I Ching*—it was a random thing. And you cannot recover the structure, it's hidden, as you said. The problem was that Stockhausen, when he played his *Klavierstück XI,* you couldn't recover the structure either. It was so complex now. So the complexity of the serialist music became exactly the complexity of Cage. Cage looked his numbers up in random number tables; the others were sitting calculating rows of numbers. But in addition to that they also had to fake it. Because—you find that yourself when you do serial music—the music moves too slowly. So you change the numbers to get the music up a little bit.

FLYNT: Yes. We're taking longer on this than I meant to ...

HENNIX: But I wanted to say this. The completely deterministic composition technique and the completely random, aleatoric technique, gave exactly the same results. And that was the complete breakdown of the Darmstadt school. That's when they started to improvise in Darmstadt. Not before that was there improvisation in Darmstadt.

FLYNT: When they first tried to serialize duration, they tried to pick a fundamental unit and use multiples of it; in other words, that's not the way you serialize pitch. You don't take one cycle per second and then use two cycles per second, up to twelve. That's not what you do. But that's what they did with duration. And that's what produced the Boulez pieces that move so slowly. In other words if you treat rhythm as multiples of like a whole note then it was moving too slowly for them.

But Cage was for them what was wrong with America or something. I mean, the center of what Stockhausen was doing was the concept of scientificity. In other words at that time I fantasized the composer appearing as performer, on the stage in a lab coat carrying a slide rule—there were no electronic calculators at that time, it would have to have been a slide rule—but that seemed completely appropriate. In other words, a composition was a laboratory experiment. I mean they viewed Cage as a typical American—coming in a vacuum—American superficiality—a vacuum with no scientificity. But Cage was actually not using a random number table, he was flipping coins, he was using the *I Ching*. Yet it was not even that—what Cage was doing was much more whimsical than using a random number book. He would just copy a leaf—in the *Concert for Piano and Orchestra* he just put the staff over a leaf and then the main points defining the shape of the leaf he just copied them on and he ended up with a circle or not a circle, but a group of notes in cyclic shape, and so the pianist was supposed to play around the circle. This was completely whimsical actually and yes, I remember very well these debates that they had, the one and the other*—I didn't have any idea that I was going to spend this much time competing with the music critic of the *New York Times* about who remembers the 1950s the best.

At any rate... There is of course a larger tradition in art which has a kind of quasi-scientific involvement in structure that does go very much to the Renaissance, for example. Althought I was not so conscious of that—I looked that up much later. But it was certainly there.

So, on the one hand concept art came from the idea of lifting structure off and making a separate art form out of it. The structure or conceptual aspect, and making a separate art form out of it. The other thing that was coming—the development of my philosophical thinking—I have to explain first that the version of mathematics that I received at Harvard in the 1950s in which Quine was the head of the department and editor of the *Journal of Symbolic Logic* and so forth and the hottest thing in philosophy was considered to be Quine's debate with Carnap. And I was a schoolmate of Kripke, Solovay, Goodman etc. etc., etc. I'm just mentioning that to locate the period of time. Actually my conversations with them were insignificant as far as the philosophy of mathematics was concerned, there was no discussion between me and them on any of that but it will locate the time frame that I'm talking

---

*Serial vs. chance.

about.*

But observing what was going on at that time, I picked up the idea that the most plausible explanation of what mathematics is, is that it is an activity analogous to chess, or in other words that chess captures the characteristic features of mathematics, even though, as I have told you privately many times, everybody knew who Brouwer was and what the intuitionist school was, but nobody studied it, and from my point of view looking at it and knowing what it was, I felt no inclination to pursue it further.

The reason why this chess game explanation of mathematics seemed so plausible—you know, at the end of the nineteenth century they found themselves with three geometries—this is not Henry Flynt saying this, this is the canard, the story in the text books. There were three geometries; one of them fit the real world. They thought it was Euclidean, but it might not be. It might be one of the others like elliptic, for example; nevertheless, all three were consistent. Now what was the epistemological status of the two out of the three geometries that were true without having any correspondence to the real world, while one of them did have a correspondence to the real world and was also true? But what of the other two—the ones that were called true even thought they had nothing to with the world? You know presumably Hilbert wrote *Foundations of Geometry* as the original answer to that question.

Although—I can't pursue this here, it is much too technical—this is now an open question for me. It has never been an open question in the past. I just accepted what I was told—that Hilbert solved this by seeing that a system of mathematics that has no relation to the real world—in what does its truth consist? Its consistency as an uninterpreted calculus as they would say—axioms, proofs, formation rules, transformation rules. Certainly it was clear in the early twentieth century that the concept of an abstract space was established. This was what geometry was about. Geometry did not attempt—in Kant's time it was assumed that when you were talking about geometry you were talking about the geometry of the real world. That's the only geometry that there was. The idea that there was a different agenda for geometry other than the real world—how Kant could have moved geometry into the constitutive subject and said that it was congenital to the mind—

---

*I'm being too diffident. I had quite significant discussions with Kripke and Goodman in 1961. [H.F., note added]

Euclidean geometry. In hindsight that seems to be one of the biggest mistakes he made, tremendously embarrassing, because by the mid-twentieth century it was completely taken for granted that the job of the mathematician was to study structures which do not have any reality. And that from time to time you will give an interpretation to one or the other of these structures, like a physical interpretation, and then it may be found to be true or false in reality or not. Meanwhile, you have another sense of the word "interpretation" which has to do with relative consistency proofs by something having a model.

This is now a completely open question for me, what they thought they were doing. In other words what Hilbert thought that he was doing—he interpreted one or another non-Euclidean geometry—what was the interpretation that he used? it was a denumerable domain of algebraic numbers [*Foundations of Geometry*, pp.27-30].

HENNIX: I think his ideas go back to Klein's models—which are Euclidean in the center of the circle and then at the periphery they have turned non-Euclidean (in the complex plane).

FLYNT: You had to have an explanation of how mathematics could be true in any sense whatsoever even though any claim of a connection with the real world had been completely severed, and it was being pursued in some kind of vacuum. What does mathematics mean in that case? And the answer that Hilbert gave was that it does not have to mean anything.

That's the answer. So it's a chess game. And the only difference between mathematics and a chess game is that there are additional complications created in mathematics by the fact that it deals with infinitary games. By the way, I completely overlooked that aspect at that time. You know, I can only see it now, kind of like two superimposed pictures, because I see what I know now and compare it with what I knew then.

HENNIX: Yeah, the same for myself. I didn't know that this idea of Hilbert's was forced by Frege until later. Frege was the one who said that either the parallel axiom is true, or it's not. Which way do you want it? And so he caused the big stir in the foundations of geometry in the end of the nineteenth century and that's why he became enemies with Hilbert. They were life enemies.

FLYNT: The reason I see it like two superimposed transparencies—

HENNIX: But even today this debate with Frege—you have to go to a single volume in Frege's posthumous writings—it is not mentioned in any textbook—no lecture mentions it, and, so far, nobody has explained it properly.*

FLYNT: Yes, yes, yes. You're talking about an obscure origin of something and what I'm talking about is a kind of consensus that had grown up, since everybody agreed that mathematics should study unreal structures.

HENNIX: But that consensus was *forced* on us, that that was what we were supposed to do.

FLYNT: The problem then—I thought mathematics was like chess. What I understand now is that even a good formalist would not agree with that. A good formalist would say that when you have a finite game like chess, the problems of validity and soundness become transparent or intuitively ascertainable, therefore a finite game is too trivial to be a proxy for mathematics. At that time I did not understand that distinction. I've read in many books since then that mathematics is the science of infinity—that is the way mathematics is defined now in half of the books that I look at. But at that point I did not understand. I thought the finite game was already, I mistakenly thought, a complex enough problem to stand for mathematics. Or that the reliability of a finite game was sufficiently complicated to stand for mathematics so I basically focused just on a finite game.

HENNIX: By the way, this was exactly the late Wittgenstein's view of the philosophy of mathematics—it's not a complete misunderstanding, that is to say, other people thought of it that way too.

FLYNT: The question then arose of even the soundness, the reliability, the consistency of a finite game—this then is the problem for example whether it is possible to follow a very simple rule correctly or not. The other thing that was feeding into everything that was going on was that Wittgenstein's *Remarks on The Foundations of Mathematics* was in the Harvard Bookstore when I walked in as a freshman my very first

---

Nachgelassene Schriften und Wissenschaftlicher Briefwechsel, vol. 2, Felix Meiner, Hamburg: 1976. (Gottlob Frege, The Philosophical and Mathematical Correspondence, University of Chicago Press: 1980)

day there—so in other words I was looking at Wittgenstein's *Remarks on The Foundations of Mathematics* from 1957—

HENNIX: Ten years before me—

FLYNT: —but very cursorily. Because I had a philosophical agenda— I passed over this material in a very cursory way because I had a philosophical agenda. I was not involved in the distinction between a finite and an infinite structure. I was not involved in that.

HENNIX: You thought there was no such distinction?

FLYNT: Well no, I thought that—it didn't seem that there was very much point in worrying about that when there were much more extreme problems to be worried about. But Wittgenstein wrote a lot about the possibility of following very simple rules. And I assumed that if there were epistemological questions for mathematics that this game interpretation—this chess interpretation—had displaced the question of the soundness and reliability of the mathematics to the possibility of understanding a very simple rule like writing the series "plus 2".

And having gathered that this was the way that I should picture mathematics—I mean we understood very well that there were other pictures of mathematics, but we thought they were philosophically obsolete. In other words the person who believed that mathematics was a description of a real supra-terrestrial structure, and certainly there were people like that—

HENNIX: Still today.

FLYNT: —we thought that this was a philosophy that had been exposed as superstitious by Positivism and possibly even by Ockham several centuries earlier. So it was not that we didn't know about that. I drew a personal conclusion that that position could not be defended by any arguments that are acceptable by modern standards. What I really meant was by Carnap's standards. That's what modern standards meant to me.

In my philosophy I was not concerned with the specifics of mathematics; I was concerned with the problem of how I know a world beyond my immediate sensations. That was actually the question that I began with—the question of propositions of material fact, like "it is raining" or "the Empire State Building is at Fifth Avenue and 34th Street."

I had read a very simplified exposition—it was actually some lectures that Carnap gave in England in the 1930s on what Positivism was.* They were very simple lectures and very different from his actual published books with all this supposed apparatus and symbols and so forth but a very simple exposition of what it is for a proposition to be meaningful—that it must be empirically testable and so forth and so on and the solution of questions of metaphysics that make assertions that are not testable are therefore meaningless—the possibility of solving questions of what is real by declaring if there is no way of deciding them they are therefore meaningless. That seemed to me to be, at the time, a stunning contribution. Because I come out of a background—I was in high school reading Kant and so forth and so on. And Carnap's solution was much more attractive to me than trying to participate with Kant, to experience his question and try to take one side or the other when he already said it's not really answerable; I solve it by simply having faith or something like that, which is what he said about the famous God freedom and immortality—I found it immensely attractive when Carnap came along and said that there is no way of answering these questions; therefore, words are being used nonsensically.

I went through a process of thinking about that without ever having seen Carnap's *The Logical Structural of The World*. When I was in Israel Scheffler's philosophy of science class, I tried to write a text which in effect gave my own empiricist constructions of what it means to say that A causes B and so forth, to give empiricist constructive definitions of those—which is, I suppose, in the spirit of Carnap's program, even though I hadn't actually seen what he had written, and if I had it would have confused me—no, I wouldn't say "confused"; I would say it would have discredited him completely. I wouldn't say "confused" because that's too modest.

HENNIX: No, I wouldn't think "confused," I would think it would have upset you ...

FLYNT: No, I wouldn't say "confused." I would say he had been discredited.

I very quickly passed to the position that the propositions of *natural science* were meaningless metaphysics.

HENNIX: On what basis? Can you pin that down? A little bit, only.

---

*R. Carnap, *Philosophy and Logical Syntax* (1935).

Philosophy of Concept Art

FLYNT: This is something I want to compress—it says a little bit about this in *Blueprint for a Higher Civilization**—like the proposition, "this key is made of iron" or something like that, I comment on that in the essay "Philosophical Aspects of Walking Through Walls."

HENNIX: I didn't recall the example actually.

FLYNT (reading): "The natural sciences must certainly be dismantled. In this connection it is appropriate to make a criticism about the logic of science as Carnap rationalized it. Carnap considered a proposition meaningful if it had any empirically verifiable proposition as an implication. But consider an appropriate ensemble of scientific propositions in good standing, and conceive of it as a conjunction of an infinite number of propositions about single events (what Carnap called protocol-sentences). Only a very small number of the latter propositions are indeed subject to verification. If we sever them from the entire conjunction, what remains is as effectively blocked from verification as the propositions which Carnap rejected as meaningless. This criticism of science is not a mere technical exercise. A scientific proposition is a fabrication which amalgamates a few trivially-testable meanings with an infinite number of untestable meanings and inveigles us to accept the whole conglomeration at once. It is apparent at the very beginning of *Philosophy and Logical Syntax* that Carnap recognized this quite clearly; but it did not occur to him to do anything about it."

The only point that I'm trying to make here is that I began to move very quickly when I was still very young towards a position of extreme disillusionment and cognitive extremism. I moved very quickly. This was not a slow process. I just immediately took Carnap's critique of metaphysics, decided that it applied directly to natural science—you dismiss natural science as meaningless. The problem: is there an object that is beyond my experience, is there a glass which is beyond what they would call the "scopic" glass, the "tactile" glass [gestures toward the glass from which he has been drinking]—is there a glass other than those glasses—when you first think about it, that question seems to have exactly the status of the propositions about God, freedom, and immortality that Kant said are unanswerable and that Carnap said are meaningless. However, there is one additional step for people who are interested in the history of philosophy. Kant, in the second edition of

---

* H. Flynt, *Blueprint for a Higher Civilization* (Milan, 1975).

Critique of Pure Reason, added this notorious refutation of idealism to prove the existence of the real world independent of my sense impressions—you may not know about this—this was the basis of Husserl's phenomenology—Husserl's phenomenology was invented in this passage and it also tremendously preoccupied Heidigger. It was one of the sources which causes Heidigger to say that the essence of Being is Time. Kant said that essentially it is the passage of time which proves that there must be an external world. This is notorious in the history of philosophy. Because on the one hand it is so deeply influential for later thinkers; and on the other hand, for example, Schopenhauer said it was a complete disgrace—it was such an obvious sophistry that it was just disgusting—that it had the effect of ruining the *Critique of Pure Reason*.

Actually this refutation of idealism is distributed throughout the *Critique of Pure Reason*, it's not in any one place—a foot note here, a preface there, another passage somewhere else. In one of the footnotes Kant makes the same point. In order to ask the question whether there is a glass beyond my sense impression of it—I cannot ask that question ...

HENNIX: Oh you mean the *ding an sich* question.

FLYNT: Well that's what Kant would have been talking about but I don't want to fit that narrowly into Kant's controlling the terms of the discussion. I'm trying to ask it as someone who has embraced Logical Positivism and is now turning around to question Logical Positivism—you see the point that I was just making there—when you say that this key is made of iron, which is Carnap's favorite example —and then a protocol sentence, for example "if I hold a magnet near this key, the key will be attracted to the magnet"—it is not clear where Carnap stands on the question whether only my sense impressions are real —just talking about this situation—only my sense impressions are real—or is there supposed to be a *substantial* key?

By the way, I don't know Carnap's work that well. I passed over these people in a very offhand way, so much so that many times I've talked to people and they've concluded in their own mind that I don't really know philosophy because I seem to have just glanced at these people—picked up one or two points—the reason for that is that I was moving so quickly to my own terminus—I only needed to see the slightest symptom from these people to know that they were spending all their time worrying about something that it was a waste of time to

Philosophy of Concept Art

worry about since it could only be a secondary issue. Here is Carnap with this key made of iron—while I'm trying to ask is there a key other than the scopic key, the tactile key *now*—since the past and the future are beyond immediate experience. I mean they cannot be cited as evidence—or whether they are evidence or not, is the same problem. Should I believe in the past and the future even though they are not immediates? Should I believe in the glass, even though what I presumably have is a scopic glass—at this very moment, a visual glass apparition, from that should I conclude a glass?

The first reaction to that question for somebody who is coming from Kant and Carnap and who does not mind how extreme his answer is—that's the key thing. In other words, if I came to a conclusion that was completely untenable as far as social circumstances—that didn't bother me at all. At first the question whether there is a real glass beyond the apparition would seem to be an unanswerable question—one of Kant's metaphysical questions—but then you think—that if you know what the question means, then there must be a realm beyond experience, because otherwise it is unclear how the question could be understandable.

From my point of view—if you want to make an issue out of semantics—this is the profound issue. What the mathematical philosophers and philosophers of mathematics were doing, talking about semantics, interpreting geometry as an algebra and algebra as a geometry—really for the purposes of relative-consistency proofs or because they found they could solve problems by using a machinery developed in another branch of mathematics by seeing these structural similarities—but to confuse that with what I thought the bona fide semantic question is: how would I understand the question whether there is a substantial glass other than the scopic glass—you know the conclusion—I can't tell you the exact breakdown—but I am talking now about the 1961 manuscript, *Philosophy Proper**—I may have already come to the conclusion at that time—that the question itself forces a yes answer. This does not mean that a proof of the existence of the external world has been given. It meant that the proposition of the existence of the external world would verify itself even if it were false!

HENNIX: I find this extremely interesting and rewarding, what you are saying now, because I never heard you say it this way before. I just want

---

*Published in *Blueprint for a Higher Civilization*.

to ask you one question before you go on: namely, I see something for the first time which I hadn't seen before—but before you go on I just want to ask you one leading question: the simple existential statement, "there is a glass on the table." You include that also in what will be doubtable here. In other words not just "there is a glass on the table" but "there exists a glass," the existential statement. I guess I wasn't very clear now.

FLYNT: No, the thing is, the approach that I'm taking doesn't break it down the way that you're talking about. Let me tell you. You may not be *sympatico* with empiricism. When you are trying to deal with philosophy at all—you have to make some allowance for the fact— you have to understand that the philosopher may be carving up problems in a way that is temperamentally alien to you.

HENNIX: Yeah ...

FLYNT: You have to understand that. This is why somebody like Carnap would read Hegel and say it's not saying anything. Actually, Hegel *is* saying something. In fact, you might go so far as to make a case that Hegel is actually rebutting Carnap, becaue if you understand what Hegel is doing you realize even more than one would realize anyway that Carnap has an untenable position—that he's sort of—that he wants what he cannot have. He has made a set of rules that does not allow him to have the thing that he demands to have. Hegel would have seen that immediately. Carnap thinks that the problem of a logic of consistency is an easy problem and a solved problem. In effect, Hegel was saying there is something very misleading in thinking that that is a solved problem. I'm trying to give you a sense of misunderstandings between philosophers that are the results of temperamental incompatibilities.

HENNIX: What you are giving me is a two-step way to skepticism. You ask a certain question—is there something beyond this perception of the glass? And you say the answer "yes" is forced on me, but then you realize this was a meaningless question.

FLYNT: No, it's the other way around.

HENNIX: Oh, okay, but here's where you have to explain in detail because here's where I miss you.

FLYNT: Let me go through the series of steps again. The series of steps was ... I'll have to do it all at the same time. You have to understand—I

Philosophy of Concept Art

don't think that you even understand what an empiricist is. It's a peculiar attitude. And one of the reasons why you have very little training in this attitude is because people who claim to be empiricists—it's always a fraud. All people who appear in public and say they are empiricists, they are all lying all of the time. The reason that they're lying is that they have this doctrine of the construction of the world from sense impressions. That is their doctrine. But they do not stay with that doctrine. And the reason why they do not stay with that doctrine is because in addition to having the doctrine of the construction of the world from sense impressions, they also want to have things like science—

HENNIX: Ethics...

FLYNT: No, not ethics—one of the characteristics of the twentieth-century philosopher was the appearance of the tough-guy philosopher who rejects all of ethics as meaningless, which Carnap certainly did and people who are close to him like A.J. Ayer—no, they did not want ethics. But they wanted science. And the problem with wanting the construction of the world from sense impressions on the one hand and wanting science on the other is that the two finally have nothing to do with each other at all—and when they said that the two were the same thing as Carnap did—he was lying—I made a hero out of Carnap—I derived some kind of positive impulse from him or something like that without—I never actually read—my serious reading of Carnap was like three or four pages of excerpts in a paperback popularization. I owned, I had in my library Carnap's so-called real books, like *Logical Foundations of Probability* and *Meaning and Necessity* and all the rest of them and I never read them.* And in hindsight that was good, because I took his slogan seriously and assumed that he meant what he said and drew the necessary consequences of it. If I had actually read his books I would have been thrust into this massive hypocrisy, and I must say stupidity, because the man did not realize that his answers were not adequate, did not realize how preposterous his constructions of the world were —

HENNIX: I would say vulgar.

---

*Again I'm being too diffident. I thoroughly studied portions of the Carnap books I owned—beginning with *The Logical Structure of Language*, which I bought while in high school [H.F., note added].

FLYNT: Yes, yes. And...what is even worse about empiricism is, in the case of somebody like Mach, not only does he want to have his sense impressions and does he want to have his science, but he wants to have science explain sense impressions! And nevertheless it was supposed to be the sense impressions that were primary, not the science. Mach is seriously telling you, I will tell you why you see a blue book—because the frequency of blue light is—and then he gives some uncountable number, I mean some number that is pragmatically infinite, or something like that. And how do you know that blue light is exactly 3.2794835 times 10 to the 15th and not one more or less—? Well, certainly not by just looking, I'll guarantee you that! You have to go into a laboratory with a few million dollars' worth of equipment or something. But that's what it is to see that the book is blue.

I'm trying to give you the sense of what it would be to be an authentic empiricist. You ask does a glass exist; an authentic empiricist would have to say that he already has a problem with that—that he has to regard that as an undefined question or statement. It's undefined, because if you are asking me if at this moment I quote unquote *have*—interesting word there, "have"—that is what our ordinary language gives us as the idiom for this.

HENNIX: Or "suffer!"

FLYNT: Yes, "have" or "suffer," that's right. I have or I suffer a scopic glass or visual glass apparition—then that is identically true. That is identically true. If you express any surprise at that, we have a problem here. I have a scopic glass. If I say I have an apparitional glass, would that be okay?—I mean from this point of view the sense impression is not open to dispute. It's meaningless to dispute it. It's an impression, an apparition—the sense impression is that for which seeming and being are identical. For the empiricist the phase of the world or range of the world for which seeming and being are identical is the sense impression. If that seems strange to you then maybe I can make it less strange by pointing out to you to make this as clear as possible—for the empiricist to say that I have an apparitional glass is to say nothing about Reality with a capital R at all! This is the so-called subjective psychological moment—although an empiricist would never say that—the reason an empiricist would never say that is that even to call it subjective is already much too strong because that implies that you can guarantee an objectivity to compare it to. And a bona fide empiricist would not agree that my sense impression is subjective—subjective in comparison to *what?*

HENNIX: So an empiricist would be a person who would not doubt whether he had a toothache or not. In other words, if he had a toothache...

FLYNT: You would regard it as being a mistake to do what? I'm not sure about the word "toothache"—if you mean that he would not doubt whether he had a toothache sensation. Whether there is an organic—in the language of medicine—whether there is an organic substrate for the toothache impression—this in a medical sense is a question of what is called hysteria or something like that

HENNIX: Suppose I have a toothache. But now I'm an empiricist so I say I'm doubting this impression. I probably don't have a toothache.

FLYNT: No, no ...

HENNIX: I have to accept the toothache?

FLYNT: No, you don't have —

HENNIX: The glass you said was—I couldn't doubt the perception of the glass. You said that was beyond doubt, in some sense, for the empiricist.

FLYNT: It would be some kind of logical mistake to think that there was anything there to be doubted.

HENNIX: Okay. And the same with the toothache.

FLYNT: Yes, yes. I mean the point is not so much that we have come into an area in which the empiricist is prepared to have faith—that would be completely missing the point. No faith is required—that's the point. The point is that it would be some kind of logical error. Once you understand what a sense impression is, the terminology of doubt does not apply to that level.

HENNIX: I see. Just that was my question.

FLYNT: The terminology of doubt does not apply to apparitions. It doesn't make sense to doubt subjective apparitions. The empiricist is already nervous when you ask does a glass exist. If you are asking whether I have a "scopic" glass, it's identically true. Wait, wait. There are already problems there. I'll come back to them. But when you say—it sounds like what you're asking me is whether the fact that I see a glass is sufficient to prove an objective glass—that sounds like ...

HENNIX: No, no, that's not what—

FLYNT: Well, ok. Most people when they say: "do you concede that there is a glass on the table—I'm sitting here looking at it," what they mean is: "do you concede that from your visual glass apparition you should conclude an objective glass, a substantial glass?" I'm taking it for granted that you know enough about philosophy to have a sense of the full weight those two words "substantial" and "objective" have in philosophy.

HENNIX: Yes.

FLYNT: That at great length is my reaction to your question about doubting "there is a glass on the table" versus doubting "there exists a glass." A bona fide empiricist would say, "Why are you asking me this?" The scopic glass is simply here for me. As far as concluding that an objective glass exists from the existence of that apparition—the traditional problem of concluding whether the apparition is a symptom of some transcendent world—I think the word "transcendent" is sometimes used in that sense in philosophy—the world beyond any sense impression—

HENNIX: This is why I used the example of the pain—because it would be senseless for me to claim that *I* can have *your* toothache!

FLYNT: Now just a minute. An empiricist—what you're really getting at— what you're sort of squeezing out of me here—I'm glad to have it squeezed out of me—I have no embarrassment about this—is that with empiricism either you must be prepared immediately to depart absolutely from the conventional world view, or else you will just plunge yourself into a quicksand of hypocrisy. When you're asking me, can I have your toothache ... A good empiricist would say, "I have not established so-called other people except the other-people apparitions that occur for me from time to time in waking life *as they do in my dreams!* And are you now going to ask me can I have the toothache of a person who appears to me in a dream?" Then the spotlight would be turned on *you*— what kind of an issue are you trying to make there? What do you believe is the reality status of the furniture in my dreams? For the empiricist, nothing remotely like that question has arisen yet, because I haven't got outside of my own quote unquote head yet.

Maybe you're just squeezing more and more. Either the empiricist must be a "madman" or else he must be insincere. I took the alternative of the madman. This is important not for me but for the general public to be told—something which the general public has never been told—

and I know why they have never been told—maybe it is necessary to complete this point. The point is that empiricism was contrived to paper over a kind of—I mean there was sort of this epistemological— Science epistemologically was resting on some sort of very shaky foundation—they saw that. They brought in this empiricism in the hope that it would solve a problem, that it would substantiate science while at the same time it would cut away the common-sense notion of causality as being unnecessary to science. Empiricism was going to give you a more sophisticated science that did not need the traditional metaphysical or common-sense notion of causality. It told you how to get along without that, but at the same time it validated everything that the scientist needed. And, at the same time, empiricism was supposed to be—in the case of Neurath—he wanted to make some kind of unification of empiricism with Marxism and make it like a complete demythified view of society.

HENNIX: There was even an attempt to bring ethics into it.

FLYNT: Well, in Neurath's case, yes.

HENNIX: Schlick too, I think—Schlick, I recall, did something in ethics.*

FLYNT: I was talking about why empiricism is not portrayed honestly in the general picture that exists of philosophy—the public picture of philosophy—it was brought in to solve the problem of what is a base for science—namely, sense impressions are going to be taken as elemental. Science is going to arise from sense impressions by construction. Nevertheless it is required that both scientific knowledge and the common-sense social world be produced by this approach—

HENNIX: Neurath, you mean.

FLYNT: No, no. Well, Carnap did not deny the existence of other people. All of the positivists ...

HENNIX: Rather, he had nothing to say about it.

FLYNT: I didn't say ethics—I said the common-sense social world. I wasn't talking about anything ethical ...

HENNIX: The existence of tables and cars and—

Fragen der Ethik, Vienna, 1930.

FLYNT: Well, what I'm saying is that the existence of other people is on the same level as the existence of tables and automobiles. And what is even worse than that is that the ones who were scientists in fact wanted to see perception itself as the product of the abstract and quantified sequence that the biophysicist or the psychophysicist sees—the light, the lens, the retina, the optic nerve, the visual cortex, and so forth and so on—they wanted to have that as prior to the sense impression but at the same time they wanted to have all that constructed up from the sense impressions. Why would this remain in place? Because it was a more palatable—it's just like why would formalism remain in place? Everybody learns that formalism died with Gödel's incompleteness theorems—it certainly didn't die for me; it isn't even clear what the incompleteness theorems are supposed to have done or not to have done—the fact remains that if you don't explain mathematics as an uninterpreted calculus, then for us there was nothing left but superstition. Those are the choices that you are given. If you don't explain that science is constructed up from a ground of sense impressions, then how do you want it to be constructed, down from God? You see, we don't take that *seriously* anymore.

As a matter of fact Hume wrote two philosophical works and in the first work* there is the notorious passage in which he himself understands what it means to be a genuine empiricist.** He says, "I feel that I am an outcast from the human race," and so forth in this famous passage—he says, "I do not know if the glass continues to exist after I've looked away from it." That line in Hume should have told you whatever you wanted to know about the existence of the glass. You should be able to ascertain the appropriate answer to your question. Hume says: "I do not know if the glass exists when I look away from it."

Hume's second book***, when he was trying to vindicate himself, when he had dropped the whole business of being a madman, it was much nearer to what empiricism means today: an attempt to construct science from a more meager inventory of elements, namely sense impressions. And that is where Hume presents his doctrine that science does not need and should not invoke metaphysical causation, that it should replace the old-fashioned causation with some sort of construc-

---

**A Treatise on Human Nature*
**Book I, Part IV, VII "Conclusion"
****An Enquiry Concerning Human Understanding*

tion which is more flat or more network-like.

Well, at any rate, I'm going into this long thing—this is why it's never dealt with in public in a sincere way—the only time it was was by the guy who invented it, Hume, in the book that he wrote when he was twenty-three years old. That's the only honest version of it and everything after that is a fraud.

The way it goes is this: I ask the question whether there is a substantial glass, an objective glass, a material glass, something that is over and above the visual glass of the moment. When first considered this seems to be a question which I have no method of answering. That would seem to place it like a Kantian metaphysical question which doesn't have a provable solution, though interestingly enough Kant thought that the existence of the external world in general could be proved but only in the second edition. And in that second edition in those little passages, Kant did really get into the existence of this individual thing like a unicorn and how that would or would not fit into the general proof of the existence of the world and also the question of how dreams would affect the validity of the proof. He touches on all of those in a way which is just awful. It's a disgraceful performance. But he had the issue there, actually.

Well, your first reaction is, "I have no way of answering this." Your second reaction is, that *if I understand the question,* then there must be an external world. So it would seem that I have actually proved the external world—that's what Kant actually said. Or he came very near to saying something like that. The third step is the realization that the statement would validate itself not only if it's true—but if it's false it validates itself equally well!

HENNIX: Given *this* method of understanding the question. And the method remained unspecified so far—as far as I know nobody has been able to do very well at specifying it.

FLYNT: What? Do you mean if somebody asks whether there is an external world—my last remark is a comment about semantics—the genuine semantic issue, as I said, and it's very different from the sort of thing that Tarski is going on about which I think is just ridiculous.

Maybe I'd better stop and tell you why I think it's ridiculous. It's because I'm now talking about things which are exactly the fundamental issues. If Tarski thinks that he can talk about the theory of chess before the question of whether the universe exists or not has been answered—they are deliberately creating specialized problems which in their minds do have answers and then they are proceeding to answer them. The larger question of whether the work has any meaning at all—it's like somebody spending his whole life working on the King's Indian defense in chess or something like that, and thinking that somehow that makes it unnecessary to answer such questions as does the chess board exist or is it only apparitional? If it's only apparitional then there is no guarantee of the continuity of the position of the pieces in the *absence* of *moves*. What happens is that people treat those basic questions as if they are so basic that it's sort of preposterous to make an issue of them. Kripke said very clearly in his book on Wittgenstein that once the question, "Does language exist?" has been asked, not to give an affirmative answer is "insane and intolerable."* It's the same reaction as there is to solipsism—that solipsism is the philosophy of the man in the lunatic assylum.

The thing that may come before all the discussion so far is the question of *what is my position on being classified as insane.* is the beginning This of philosophy for me.

HENNIX: Well, this is the classical beginning of philosophy.

FLYNT: Because if you're not willing to face up to being classified as insane—if you want to avoid that confrontation—you can't be a philosopher. That confrontation is at the center of bona fide philosophy.

HENNIX: Or was ...

FLYNT: Yes. At any rate, I had reached this point in something like 1961. I had not yet done the "Is there language?" trap. But I had reached the point of saying that to claim the existence of a world beyond experience is untenable. However I understood very well that it begins to create problems for me to say, I have a visual glass apparition, because there is a lot of structure in that sentence. And it's not clear what is supporting that structure after the world has been cut away. Even the use of the idioms like "have" and "suffer." The use of the word "I"—after the objective world has been cut away it's unclear what is the basis for all of that. And this is the point I had reached in 1961 and this is the point when I did Concept Art.

On the one hand you have an art which is about structure and

---

*S. Kripke, *Wittgenstein on Rules and Private Language*, p.60

Philosophy of Concept Art

conceptual things. On the other hand this art is not going to *affirm* traditional doctrines of structuredness and conceptualization. It is deliberately in every case going to violate them. It is going to express the fact that there has been a philosophical discovery made. I would have said chess is not a sound game. It's not well founded. It can't be. The whole problem of Wittgenstein's famous question—what is the meaning of a rule? My answer would be it doesn't have one. When you look at it from the standpoint of Hume when he says I have become a monster, I am outside the human race—the standpoint of the person who chooses insanity as opposed to intellectual dishonesty!

The person who chooses being a madman—even chess doesn't work. The whole question of its consistency. The point of Concept Art is on the one hand to transmit the tradition from the isorhythmic motet and the five Platonic solids, in Leonardo—and on the other it's to blow it up because each work of concept art must be a counter-example to that tradition. And at the same time to say that it is art means—when I passed to Concept Art I left behind many things that traditionally would have been considered crucial features of art, like sentiment, for example. Let me just leave it at that.

When the Renaissance people did study geometry and art, they developed perspective to paint people, not to paint abstractions. And you know I have to admit quite bluntly, my Concept Art was already the product of the acceptance of an abstract art. And now, many years later I can see that that was an historical juncture, to consider it tolerable that art should break with sentiment and with the representation of people. It's like moving toward an Islamic view of art. And then saying, now however, in the future, instead of Mosque decoration we will do a piece that has the visual, sensuous delectation, but it's completely abstract. But whereas Islamic art was trying to express the *truth* of a certain theorem in group theory, Concept Art must express that you can't have that—that that theorem fails. Now I'm formulating an unsolved problem—I never did a concept piece the purpose of which was to rebut the symmetry involved in a visual pattern, with that as the opponent to be hit. I mean I very well could and perhaps should.

All of my pieces were uninterpreted calculi. Because I accepted that that was the only way of explaining what mathematics is: that it consists of a body of truth about a world that does *not* exist, and explicitly so. And that all of the traditional explanations of mathematical content are now seen to be anachronistic superstitions. They are just indefensible in the modern world. Put those two things together and mathematics becomes a chess game, an uninterpreted calculus.

All of my Concept pieces are using the terminology of Carnap's *Logical Syntax of Language*—the formation rule, the transformation rule—but in each case they wish to express the violation, the failure of some traditional organizing principle of these uninterpreted calculi. For instance there is one where, among other things, the very notation itself has an undisplaced active interaction with the subjectivity of the quote unquote reader.* And that determines the structure of the derivation, the proof. It was pointed out to me many years later that it's not just that you don't get this in schoolbook mathematics—this is what they are *most concerned to exclude*.

I had another one, in which there was no general transformation rule.** There were only completely nominalistic transformation rules. In other words, for each step you are told, for that step only and for this moment only, what the transformation rule is. And by the time you are ready to take the next step, that rule is forgotten and inoperative.

HENNIX: This is the Energy Cube Organism?

FLYNT: No, no. The Energy Cube Organism was not Concept Art at all. No, no. It was a different genre. That one was the piece called "Transformations."

The Energy Cube Organism and the Perception-Dissociator in my own classification are not Concept Art. Only the pieces labeled "Concept Art" are Concept Art. And I only did four of them until 1987. Three of them are in *An Anthology*, and the fourth was published in *dimension* 14 (1963). The Energy Cube Organism and the Perception-Dissociator were in other genres. I drew these distinctions of genre rather narrowly, actually.

This is the one [pointing to 6/19/61 in *An Anthology*] where there is, in an uninterpreted calculus, interaction between notation and the subjectivity of the quote unquote reader.

This is "Transformations." You are just taking these objects, you are burning them, melting them, doing all sorts of things to them. The point of this is that each step in the proof—you have to think of it as a proof—you see it has the tree structure of a proof. This is my nominalistic transformation rules, because each rule is stipulated only at that step, and then it is thrown away. The point that I was trying to express

---

*dated 6/19/61—later titled "Illusions."
** "Transformations," retitled "Implications" in the second edition.

Philosophy of Concept Art

was that's what they do in all of it—even in chess, when you move the pawn to King's Bishop 3, you think that you are conforming to a general rule written in Heaven. But in fact there isn't any general rule, and when you move the pawn to Bishop 3, you're just making up what you are doing right at that moment, and there isn't any general rule.

HENNIX: You would label this ad hoc?

FLYNT: That's right. That would be perhaps a better word for it. All transformation rules and probably even all formation rules are ad hoc, yes, yes.

I said "nominalistic" because they are only there individually. They do not add up to any general—

HENNIX: System of rules?

FLYNT: No—not that—they do not add up to any generality, to a general rule that covers all cases of a certain class.

What is inadequate about this—and I realized very quickly that it's inadequate—is that this does not actually give some profound reason in concrete practice for questioning chess. That's what the inadequacy of the original Concept Art pieces is. That they don't really give you some kind of operative situation where you can see that following the chess rules is failing. I don't provide that. I only provide something that's ritualistic. Saying this is how you would behave if you realized that following any rule is ad hoc.

A conventional mathematician would say, you have not proved that the world that this is designed for is the world that I have to live in. THAT's the inadequacy. He would say that I am only ritualizing the world of impoverishment or disorganization. I'm not showing that that's the world that people in general have to live in because it's in force. That's the difference between then and now. The reason that I want meta-technology would be to give a situation where somebody can actually see that you CAN'T play a game of chess—or that you want to play one and that I, by putting it in the appropriate context, make it clear that the general rules on which playing it depends are not in fact available.

But to show that in a serious way. From the prevailing point of view I would be talking about contriving a miracle. In other words, to actually substantiate any of these—what is interesting is not so much "Transformations"—but it would be some situation that would substantiate that the conventional view is actually unavailable. And to do

that you have to violate what are considered today to be the soundest laws of science. I'd need a miracle to manifest that I'm right, so to speak. So by the time I get to meta-technology I'm in the job of constructing miracles, I mean constructing situations that are absolutely physically impossible (or in some cases logically impossible) by currently accepted scientific and commonsense views of what is the real world.

"Innperseqs" is the one that is visually sensuously the best. You are making a rainbow halo that you can get by breathing on your glasses and looking at a point light—you get a rainbow halo around the light. Eventually I will set it up so that you don't need glasses or anything so that the whole business of seeing the rainbow halo is moved out and does not require any special preparation by the spectator. The rainbow halo is the sensuous delectation. The derivation, the proof, the specification of propositions, is something that you do as the halo is fading. You have to quickly specify—I never analyzed exactly what was going on there but it was as if—you have a notation which is externally changing, and therefore the quote unquote reading of a mathematical system has to be a process that is taking place in experienced time.

By acts of attention you have to choose sentences, to choose implications—it's a display. You are given an external display which is changing out there, not in your head. And you have to place a structure on it by specified rules.

You know another point that can be made is, that "Innperseqs" is philosophically inconsistent with "Transformations"—that these pieces are mocking each other.

At the time that I did this, I did not have the kind of maturity that I would have today to put it together in a strong way. These were gestures. And they are not even uniform on a question like whether a rule exists or not. Well actually, frequently I'm too hard on myself. I think that in the essay "Concept Art" I do say something like, objective language doesn't exist, but I'm still free to work with what you think the text says—I can use that in *art:* this is *art!*

There are three ways that the art part comes in. One is the visual display, the delectation. The second way the art part comes in is—well, if LaMonte Young's Word Pieces are art, then this is art too. But the third thing is that this does not claim to have objective truth. It is a construction for the world-hallucination or the world-apparition or even a construction for the private world-apparition.

Philosophy of Concept Art

HENNIX: You are actually extending the world by new constructions.

FLYNT: But it's the world-apparition. In a sense if I believed that these rules were objectively established, then it would almost indicate that I had not learned the lesson of the very piece which sits beside it on the page!* And what am I doing talking about a page and a text? So the answer is that I have abandoned the provision of truth as the purpose of this activity and I have moved to the provision of experiences where the possibility of these experiences is a surprise.

HENNIX: And you don't have to be an empiricist to be surprised.

FLYNT: Yes. Yes. But the truth claim that you would have from a Kripke or a Goodman has been dropped. The meaning of the text is the meaning that the reader associates to it. And the thing is, that in conventional intellectual work that's an unacceptable answer, because usually you are trying to get independent of the reader's distortion—that's the whole hope—that you can make something that is independent of the reader's distortion of it. This is a different game. This is not classical mathematics; it's not classical science. It's like giving a Rorschach blot. Then I don't mind if you have a unique subjective reaction. If my purpose is to make Rorschach blots, then I do not object, I have not failed, if you have a unique personal reaction.

These pieces are designed for the individual reaction rather than in spite of it.

The only other Concept Art piece—in *dimension* 14—"one just has to guess whether this piece exists and if it does what its definition is." That was the piece. And that was a response to Cage's dissociation of what the composer sees, the performer sees, the audience sees. Starting from that, going through all the games that LaMonte had played with the idea of performance, where we were performing pieces first and composing them second, maybe many months later. So finally with the Concept Art piece, even whether the piece exists is completely indeterminate, but I meant for people to try to take that seriously. I was having a joke with the person who thinks that concepts form an objective world, which the individual who cognizes only discovers bit by bit. In effect, I am giving him this: thank you for believing that there is a piece here—I'm leaving it to you to find it. I wash my hands of that problem—*you* find it!

---

*"Innperseqs" versus "Transformations," second edition.

Well, there's a natural pause that comes here because I think that I've summarized perhaps fairly thoroughly where I was when I did the work published in 1963. The entire subsequent career of the label Concept Art, its misapplication to Word Pieces and all the rest of it, we have not begun with. After that, we can go on to the discussion of your visual pieces of the 70s and how they resume the genre of Concept Art.

(End of Part I)

Concept Art

Henry Flynt

On the following pages appears a facsimile of Henry Flynt's 1961 "Concept Art" manuscript as published in *An Anthology,* ed. La Monte Young (New York, 1963). Typographical errors are of course reproduced; the most serious is the misspelling of "intension" in the first paragraph. The manuscripts by Flynt cited in the essay were subsequently published, as noted below. The piece dated 6/19/61 was subsequently titled "Illusions." An enlargement of the text "Transformations" was hung at "The Arts in Fusion" exhibition, Tyler School of Art, Temple University, January 1966—the first exhibition of a work designated as concept art. The other early concept art piece to be published was "Work Such That No One Knows What's Going On," in *dimension 14* (Ann Arbor, 1963).

"Philosophy Proper" and 1966 Mathematical Studies" appear in H. Flynt, *Blueprint for a Higher Civilization* (Milan, 1975).
"Structure Art and Pure Mathematics" (1960) appears in Henry Flynt, *Fragments & Reconstructions from a Destroyed Oeuvre, 1959-1963* (New York, Backworks, 1982).

ESSAY: CONCEPT ART   (PROVISIONAL VERSION)

"Concept art" is first of all an art of which the material is "concepts", as the material of for ex. music is sound. Since "concepts" are closely bound up with language, concept art is a kind of art of which the material is language. That is, unlike for ex. a work of music, in which the music proper (as opposed to notation, analysis, a.s.f.) is just sound, concept art proper will involve language. From the philosophy of language, we learn that a "concept" may as well be thought of as the <u>intesion</u> of <u>a name</u>; this is the relation between concepts and language. The notion of a concept is a vestige of the notion of a platonic form (the thing for which for ex. all tables have in common: tableness), which notion is replaced by the notion of a name objectively, metaphysically related to its intension (so that all tables now have in common their objective relation to 'table'). Now the claim that there can be an objective relation between a name and its intension is wrong, and (the word) 'concept', as commonly used now, can be discredited (see my book, <u>Philosophy Proper</u>). If, however, it is enough for one that there be a subjective relation between a name and its intension, namely the unhesitant decision as to the way one wants to use the name, the unhesitant decisions to affirm the names of some things but not others, then 'concept' is valid language, and concept art has a philosophically valid basis.

Now what is artistic, aesthetic, about a work which is a body of concepts? This question can best be answered by telling where concept art came from; I developed it in an attempt to straighten out certain traditional activities generally regarded as aesthetic. The first of these is "structure art", music, visual art, a.s.f., in which the important thing is <u>"structure"</u>. My definitive discussion of structure art can be found in "General Aesthetics"; here I will just summarize that discussion. Much structure art is a vestige of the time when for ex. music was believed to be knowledge, a science, which had important things to say in astronomy a.s.f.. Contemporary structure artists, on the other hand, tend to claim the kind of cognitive value for their art that conventional contemporary mathematicians claim for mathematics. Modern examples of structure art are the fugue and total serial music. These examples illustrate the important division of structure art into two kinds according to how the structure is appreciated. In the case of a fugue, one is aware of its structure <u>in listening to it;</u> one imposes "relationships", a categorization (hopefully that intended by the composer) on the sounds while listening to them, that is, has an "(associated) artistic structure experience". In the case of total serial music, the structure is such that this cannot be done; one just has to read an "analysis" of the music, definition of the relationships. Now there are two things wrong with structure art. First, its cognitive pretensions are utterly wrong. Secondly, by trying to be music or whatever (which have nothing to do with knowledge), and knowledge represented by structure, structure art both fails, is completely boring, as music, and doesn't begin to explore the aesthetic possibilities structure can have when freed from trying to be music or whatever. The first step in straightening out for ex. structure music is to stop calling it "music", and start saying that the sound is used only to carry the structure and that the real point is the structure--and then you will see how limited, impoverished, the structure is. Incidentally, anyone who says that works of structure music do occasionally have musical value just doesn't know how good real music (the Goli Dance of the Baoule; "Cans on Windows" by L. Young; the contemporary American hit song "Sweets for My Sweets", by the Drifters) can get. When you make the change, then since structures are concepts, you have concept art. Incidentally, there is another, less important kind of art which when straightened out becomes concept art: art involving play with the concepts of the art :such as, in music, "the score", "performer vs. listener", "playing a work". The second criticism of structure art applies, with the necessary changes, to this art.

The second main antecedent of structure art is mathematics. This is the result of my revolution in mathematics, which is written up definitively in the appendix; here I will only summarize. The revolution occured first because for reasons of taste I wanted to de-emphasize discovery in mathematics, mathematics as discovering theorems and proofs. I wasn't good at such discovery, and it bored me. The first way I thought of to de-emphasize discovery came not later than Summer, 1960; it was that since the value of pure mathematics is now regarded as aesthetic rather than cognitive, why not try to make up aesthetic theorems, without considering whether they are true. The second way, which came at about the same time, was to find, as a philosopher, that the conventional claim that theorems and proofs are discovered is wrong, for the same reason I have all ready given that 'concept' can be discredited. The third way, which came in the fall-winter of 1960, was to work in unexplored regions of formalist mathematics. The resulting mathematics still had statements, theorems, proofs, but the latter weren't discovered in the way they traditionally were. Now exploration of the wider possibilities of mathematics as revolutionized by me tends to lead beyond what it makes sense to call "mathematics"; the category of "mathematics", a vestige of Platonism, is an "unnatural", bad one. My work in mathematics leads to the new category of "concept art", of which straightened out traditional mathematics (mathematics as discovery) is an untypical, small but intensively developed part.

I can now return to the question of why concept art is "art." Why isn't it an absolutely new, or at least a non-artistic, non-aesthetic activity? The answer is that the antecedents of concept art are commonly regarded as artistic, aesthetic activities; on a deeper level, interesting concepts, concepts enjoyable in themselves, especially as they occur in mathematics, are commonly said to "have beauty". By calling my activity "art", therefore, I am simply recognizing this common usage, and the origin of the activity in structure art and mathematics. However: it is confusing to call things as irrelevant as the emotional enjoyment of (real) music, and the intellectual enjoyment of concepts, the same kind of enjoyment. Since concept art includes almost everything ever said to be "music", at least, which is not music for the emotions, perhaps it would be better to restrict 'art' to apply to art for the emotions, and recognize my activity as an independent, new activity, irrelevant to art (and knowledge).

Transformations -   Concept Art Version of Colored Sheet Music No.1   3/14/61 (10/11/61)
The initial object: a sheet of cheap, thin white typewriter paper
Transformation of the initial obj. (obj.1) into obj. 2: soak the initial obj. in inflammable liquid which does not leave solid residue when burned; then burn it on horizontal rectangular white fireproof surface - obj. 2 is ashes (on surface)
Transformation of object 2 into obj. 3: make black and white photograph of obj. 2 in white light (image of ashes' "rectangle" with respect to white surface (that is, of the region (of surface, with the ashes on it) with bounding edges parallel to the edges of the surface and intersecting the four points in the ashes nearest the four edges of the surface) must exactly cover the film); develop film - obj. 3 is the negative

Transformation of obj. 2 and obj. 3 into obj.4: melt obj. 3 and cool in mold to form plastic doubly convex lens with small curvature; take color photograph of ashes' rectangle in yellow light using this lens; develop film – obj.4 is color negative

Transformation of obj.2 and obj.4 into obj.5: repeat last transformation with obj. 4 (instead of 3), using red light – obj. 5 is second color negative

Transformation of obj. 2 and obj. 5 into obj.6: repeat last transformation with obj. 5, using blue light – obj. 6 is third color negative

Transformation of obj.2 and obj.6 into obj.7: make lens from obj. 6 mixed with the ashes which have been being photographed; make black and white photograph, in white light, of that part of the white surface where the ashes' rectangle was; develop film – obj.7 is second black and white negative

Transformation of obj. 2, obj. 6, and obj. 7 into the final obj. (obj. 8): melt, mold, and cool lens used in last transformation to form negative, and make lens from obj.7 ; using negative and lens in an enlarger, make two prints, an enlargement and a reduction – enlargement and reduction together constitute the final object

Concept Art Version of Mathematics System 3/26/61(6/19/61)
An "element" is the facing page (with the figure on it) so long as the apparent, perceived, ratio of the length of the vertical line to that of the horizontal line (the element's "associated ratio") does not change.

A "selection sequence" is a sequence of elements of which the first is the one having the greatest associated ratio, and each of the others has the associated ratio next smaller than that of the preceding one. (To decrease the ratio, come to see the vertical line as shorter, relative to the horizontal line, one might try measuring the lines with a ruler to convince oneself that the vertical one is not longer than the other, and then trying to see the lines as equal in length; constructing similar figures with a variety of real (measured) ratios and practicing judging these ratios; and so forth.) [Observe that the order of elements in a selection sequence may not be the order in which one sees them.]

Concept Art: Innperseqs (May – July 1961)

A "hălpoint" iff whatever is at any point in space, in the fading rainbow halo which appears to surround a small bright light when one looks at it through glasses fogged by having been breathed on, for as long as the point is in the halo.

An "init′point" iff a halpoint in the initial vague outer ring of its halo.

An "inn′persĕq" iff a sequence of sequences of halpoints such that all the halpoints are on one (initial) radius of a halo; the members of the first sequence are initpoints; for each of the other sequences, the first member (a "consequent") is got from the non-first members of the preceding sequence (the "antecedents") by being the inner endpoint of the radial segment in the vague outer ring when they are on the segment, and the other members (if any) are initpoints or first members of preceding sequences; all first members of sequences other than the last appear as non-first members, and halpoints appear only once as non-first members; and the last sequence has one member.

Indeterminacy

A ⌜totally determinate innperseq⌝ iff an innperseq in which one is aware of (specifies) all halpoints.

An ⌜antecedentally indeterminate innperseq⌝ iff an innperseq in which one is aware of (specifies) only each consequent and the radial segment beyond it.

A ⌜halpointally indeterminate innperseq⌝ iff an innperseq in which one is aware of (specifies) only the radial segment in the vague outer ring, and its inner endpoint, as it progresses inward.

Copyright by Henry A. Flynt Jr., 1961

The Apprehension of Plurality

Henry Flynt

(An instruction manual for 1987 concept art)

I. Original Stroke-Numerals

Stroke-numerals were introduced in foundations of mathematics by the German mathematician David Hilbert early in the twentieth century. Instead of a given Arabic numeral such as '6', for example, one has the expression consisting of six concatenated occurrences of the stroke, e.g. 'IIIIII'.

To explain the use of stroke-numerals, and to provide a background for my innovations, some historical remarks about the philosophy of mathematics are necessary. Traditional mathematics had treated positive whole-number arithmetic as if the positive whole numbers (and geometrical figures also) were objective intangible beings. Plato is usually named as the originator of this view. Actually, there is a scholarly controversy over the degree to which Plato espoused the doctrine of Forms—over whether Aristotle's *Metaphysics* put words in Plato's mouth—but that is not important for my purposes. For an intimation of the objective intangible reality of mathematical objects in Plato's own words, see the remarks about "divine" geometric figures in Plato's "Philebus." Aristotle's *Metaphysics,* I.6, says that mathematical entities

> are intermediate, differing from things perceived in being eternal and unchanging, and differing from the Forms in that they exist in copies, whereas each Form is unique.

For early modern philosophers such as Hume and Mill, any such "Platonic" view was not credible and could not be defended seriously. Thus, attempts were made to explain number and arithmetic in ways which did not require a realm of objective intangible beings. In fact, Hume said that arithmetic consisted of tautologies; Mill that it consisted of truths of experience.

Following upon subsequent developments—the philosophical climate at the end of the nineteenth century, and specifically mathematical developments such as non-Euclidian geometry—Hilbert proposed that mathematics should be understood as a game played with meaningless marks. So, for example, arithmetic concerns nothing but formal terms—numerals—in a network of rules. Actually, what made arithmetic problematic for mathematicians was its infinitary character—as expressed, for example, by the principle of complete induction. Thus, the principal concern for Hilbert was that this formal game should not, as a result of being infinitary, allow the deduction of both a proposition and its negation, or of such a proposition as $0 = 1$.

But at the same time (without delving into Hilbert's distinction between mathematics and metamathematics), the stroke-numerals replace the traditional answer to the question of what a number is. The stroke-numeral 'IIIIII' is a concrete semantics for the sign '6', and at the same time can serve as a sign in place of '6'. The problem of positive whole numbers as abstract beings is supposedly avoided by inventing e.g. a number-sign, a numeral, for six, which is identically a concrete semantics for six. Let me elaborate a little further. A string of six copies of a token having no internal structure is used as the numeral '6', the sign for six. Thus the numeral is itself a collection which supposedly demands a count of six, thereby showing its meaning. Hans Freudenthal calls this device an "ostensive numeral."

So traditionally, there is a question as to what domain of beings the propositions of arithmetic refer to, a question as to what the referents of number-words are. *Correlative to this, mathematicians' intentions require numerous presuppositions about content, and require extensive competancies—which the rationalizations for mathematics today are unable to acknowledge, much less to defend.*

For example, if mathematics rests on concrete signs, as Hilbert proposed, then, since concrete signs are objects of perception, the reliability of mathematics would depend on the reliability of perception. Given the script numeral **1**

The Apprehension of Plurality

which is ambiguous between one and two, conventional mathematics would have to guarantee the exclusion of any such ambiguity as this. Yet foundations of mathematics excludes perception and the reliability of concrete signs as topics—much as Plato divorced mathematics from these topics. (Roughly, modern mathematicians would say that reliability of concrete signs does not interact with any advanced mathematical results. So this precondition can simply be transferred from the requisites of cognition in general. But it would not be sincere for Hilbert to give this answer. Moreover, my purpose is to investigate the possibility of reconstructing our intuitions of quantity beyond the limits of the present culture. In this connection, I need to activate the role of perception of signs.)

But the most characteristic repressed presuppositions of mathematics run in the opposite, supra-terrestrial direction. Mathematicians' intentions require a realm of abstract beings. Again, it is academically taboo today to expose such presuppositions.* But to recur to the purpose of this investigation, concept art is about reconstructing our intuitions of quantity beyond the limits of the present culture. This project demands an account of these repressed presuppositions. To compile such an account is a substantial task; I focus on it in a collateral manuscript entitled "The Repressed Content-Requirements of Mathematics." To uncover the repressed presuppositions, a combination of approaches is required.** I will not dwell further on the matter here—but a suitable sample of my results is the section "The Reality-Character of Pure Whole Numbers and Euclidian Figures" in "The Repressed Content-Requirements."

Returning to the original stroke-numerals, they were meant (among other things) to be part of an attempt to explain arithmetic without requiring numbers as abstract beings. They were meant as signs, for numbers, which are identically their own concrete semantics. Whether I think Hilbert succeeded in dispensing with abstract entities is not the point here. I am interested in how far the exercise of positing

---

*Gödel and Quine admit the need to assume the non-spatial, abstract existence of classes. But they cannot elaborate this admission; they cannot provide a supporting metaphysics.

**One anthropologist has written about "the locus of mathematical reality"—but, being an academic, he merely reproduces a stock answer outside his field (namely that the shape of mathematics is dictated by the physiology of the brain).

stroke-numerals as primitives can be elaborated. My notions of the original stroke-numerals are adapted from Hilbert, Weyl, Markov, Kneebone, and Freudenthal. For example, how does one test two stroke-numerals for equality? To give the answer that "you count the strokes, first in one numeral and then in the other," is not in the spirit of the exercise. For if that is the answer, then that means that you have a competency, "counting," which must remain a complete mystery to foundations of mathematics. What one wants to say, rather, is that you test equality of stroke-numerals by "cross-tallying": by e.g. deleting strokes alternately from the two numerals and finding if there is a remainder from one of the numerals. This is also the test of whether one numeral precedes the other. So, now, given an adult mastery of quality and abstraction, you can identify stroke-numerals without being able to "count."

In the same vein, you add two stroke-numerals by copying the second to the right of the first. You subtract a shorter numeral from a longer numeral by using the shorter numeral to tally deletion of strokes from the longer numeral. You multiply two stroke-numerals by copying the second as many times as there are strokes in the first: that is, by using the strokes of the first to tally the copying of the second numeral.

To say that all this is superfluous, because we already acquired these "skills" as a child, misses the point. The child does not face the question, posed in the Western tradition, of whether we can avoid positing whole numbers as abstract beings. To weaken the requirements of arithmetic to the point that somebody with an adult mastery of quality and abstraction can do feasible arithmetic "blindly"—i.e. without being able to "count," and without being able to see number-names ('five', 'seven', etc.) in concrete pluralities—is a notable exercise, one that correlates culturally with positivism and with the machine age.

To reiterate, the stroke-numeral is meant to replace numbers as abstract beings by providing number-signs which are their own concrete semantics. Freudenthal said that we should communicate positive whole numbers to alien species by broadcasting stroke-numerals to them (in the form of time-series of beeps). Still, Freudenthal said that the aliens would have to resemble us psychologically to get the point. (*Lincos,* pp. 14-15.)

When Hilbert first announced stroke-numerals, certain difficulties were pointed out immediately. It is not feasible to write the stroke-numerals for very large integers. (And yet, if it is feasible to write the stroke-numeral for the integer n, then there is no apparent reason why it would not also be feasible to write the stroke-numeral for n+1. So stroke-numerals are closed under succession, and yet are contained in a finite segment of the classical natural number series.) Moreover, large feasible stroke-numerals, such as that for 10,001, are not surveyable.

But this is not a study of metamathematical stroke-numerals. And I do not wish to go into Hilbert's question of the consistency of arithmetic as an infinitary game here; "The Repressed Content-Requirements" will have more to say on the consistency question. The purpose of this manual, and of the artworks which it accompanies, is to establish apprehensions of plurality beyond the limits of traditional civilizations (beyond the limits of Freudenthal's "us"). Moreover, these apprehensions of plurality are meant to violate the repressed presuppositions of mathematics. I refer back to original stroke-numerals because certain devices which I will use in assembling my novelties cannot be supposed to be intuitively comprehensible—certainly not to the traditionally-indoctrinated reader—and will more likely be understood if I mention that they are adaptations of features of original stroke-numerals. Let me mention one point right away. In our culture, we usually see numerals as positional notations—e.g. 111 is decimal $1 \times 10^2 + 1 \times 10^1 + 1$ or binary $1 \times 2^2 + 1 \times 2^1 + 1$. But stroke-numerals are not a positional notation (except trivially for base 1). Likewise, my novelties will not be positional notations; I will even nullify the reference to base 1. (Only much later in my investigations, when broad scope becomes important, will I use positional notation.) So the foregoing introduction to stroke-numerals has only the purpose of motivating my novelties. And references to the academic canon are given only for completeness. They cannot be norms for what I am "permitted" to posit.

II. Simple Necker-Cube Numerals

In my stroke-numerals, the printed figure, instead of being a stroke, is a Necker cube. (Refer to the attached reproduction, "Stroke-Numeral.") A Necker cube is a two-dimensional representation of a cubical frame, formed without foreshortening so that its perspective is perceptually equivocal or multistable. The Necker cube can be seen as flat, as slanting down from a central facet like a gem, etc.; but for the moment I am exclusively concerned with the two easiest variants in which it is seen as an ordinary cube, either projecting up toward the front or down toward the front.

STROKE-NUMERAL

◧ STROKE
◨ VACANT

Since I will use perceptually multistable figures as notations, I need a terminology for distinctions which do not arise relative to conventional notation. I call the ink-shape on paper a *figure*. I call the stable apparition which one sees in a moment—which has imputed perspective—the *image*.* As you gaze at the figure, the image changes from one orientation to the other, according to intricate subjective circumstances. It changes spontaneously; also, you can change it voluntarily.

Strictly—and very importantly—it is the image which in this context becomes the notation. Thus, I will work with notations which are not ink-shapes and are not on a page. They arise as active interactions of awareness with an "external" or "material" print-shape or object.

So far, then, we have images—partly subjective, pseudo-solid shapes. I now stipulate an alphabetic role for the two orientations in question. The up orientation is a *stroke;* the down orientation is called *"vacant,"* and acts as the proofreaders' symbol ⌒ , meaning "close up space." (So that "vacant" is not "even" an alphabetic space.) Now the two images in question are *signs*. The transition from image to sign can be analogized to the stipulation that circles of a certain size are (occurances of) the letter "o."** I may say that one sees the image; one apprehends the image as sign.

When a few additional explanations are made, then the signs become plurality-names or "numerals." First, figures, Necker cubes, are concatenated. When this is done, a *display* results. So the stroke-numeral in the artwork, as an assembly of marks on a surface, is a display of nine Necker cubes. An *image-row* occurs when one looks at the display and sees nine subjectively oriented cubes, for just so long as

---

*I may note, without wanting to be precious, that a bar does not count as a Hilbert stroke unless it is vertical relative to its reader.

**And—the shape, bar, positioned vertically relative to its reader, is the symbol, Hilbert stroke.

The Apprehension of Plurality

the apparition is stable (no cube reverses orientation). I chose nine Necker cubes as an extreme limit of what one can apprehend in a fixed field of vision. (So one must view the painting from several meters away, at least.) The reader is encouraged to make shorter displays for practice. Incidentally, if one printed a stroke-numeral so long that one could only apprehend it serially, by shifting one's visual field, it would be doubtful that it was well-defined. (Or it would incorporate a feature which I do not provide for.) The universe of pluralities which can be represented by these stroke-numerals is "small." My first goal is to establish "subjectified" stroke-numerals at all. They don't need to be large.

The concatenated signs which you apprehend in a moment of looking at the display are now apprehended or judged as a plurality-name, a *numeral*. At the level where you apprehend signs (which, remember, are alphabetized, partly subjective images, not figures), the apparition is disambiguated. Thus I can explain this step of judging the signs as plurality-names by using fixed notation. For nine Necker cubes with the assigned syntactical role, you might apprehend such permutations of signs as

a) I◌◌II◌◌◌I

b) I◌◌◌◌◌III

c) IIII◌◌◌◌◌

d) IIII◌◌◌◌I

e) ◌◌◌◌◌◌◌◌◌

My Necker-cube stroke-numerals are something new; but (a)-(e) are not—they are just a redundant version of Hilbert stroke-numerals (which nullifies the base 1 reference as I promised). The "close up space" signs function as stated; and the numeral concluded from the expression corresponds to the number of strokes; i.e. the net result is the Hilbert stroke-numeral having the presented number of strokes. So (a) and (b) and (c) all amount to IIII. (d) amounts to IIIII.

As for (e), it has the alphabetic role of a blank. My initial interpretation of this blank is "no numeral present." Later I may interpret the blank as "zero," so that every possibility will be a numeral. Let me explain further. Even when I will interpret the blank as "zero," it will not come about from having nine zeros mapped to one zero (like a sum of zeros). (e) has nine occurrences of "close up space," making a blank.

There is always only one way of getting "blank." (A two-place display allows two ways of getting "one" and one way of getting "two"; etc.) The notation is not positional. It is immaterial whether one "focuses" starting at the left or at the right.

Relative to the heuristic numerals (a)-(e), you may judge the intended numerals by counting strokes, using your naive competency in counting. (It is also possible to use such numerals as (a)-(e) "blindly" as explained earlier. This might mean that there would be no recognition of particular numbers as gestalts; identity of numbers would be handled entirely by cross-tallying.) The Necker-cube numerals, however, pertain to a realm which is in flux because it is coupled to subjectivity. My numerals provide plurality-names and models of that realm. Thus, the issue of what you do when you conclude a numeral from a sign in perception is not simple. *We have to consider different hermeneutics for the numerals—and the ramifications of those hermeneutics.* Here we begin to get a perspective of the mutability which my devices render manageable.

For one thing, given a (stable) image-row, and thus a sign-row, you can indeed use your naive arithmetical competency to count strokes, and so conclude the appropriate numeral. This is *bicultural hermeneutic,* because you are using the old numbers to read a new notation for which they were not intended. We use the same traditional counting, of course, to speak of the number of figures in a display.

(This prescription of a hermeneutic is not entirely straightforward. The competency called counting is required in traditional mathematics. But such counting is already paradoxical "phenomenologically." I explain this in the section called "Phenomenology of Counting" in "The Repressed Content-Requirements." As for the Necker-cube numerals, the elements counted are not intended in a way which supports the being of numbers as eternally self-identical. So the Necker-cube numerals might resonate with the phenomenological paradoxes of ordinary counting. The meaning of ordinary numbering, invoked in this context, might begin to dissolve. But I mention this only to hint at later elaborations. At this stage, it is proper to recall one's inculcated school-counting; and to suppose that e.g. the number of figures in a display is fixed in the ordinary way.)

Then, there is the *ostensive hermeneutic.* Recall that I explained Hilbert stroke-numerals as signs which identically provide a concrete semantics for themselves; and as an attempt to do arithmetic without assuming that one already possesses arithmetic in the form of com-

The Apprehension of Plurality

petency in counting, or of seeing number-names in pluralities. My intention was to prepare the reader for features to be explained now. On the other hand, at present we drop the notion of handling identity of numerals by cross-tallying.* For the ostensive hermeneutic, it is crucial that the display is short enough to be apprehended in a fixed field of vision.

With respect to short Hilbert numerals, I ask that when you see e.g.

||

marked on a wall, you grasp it as a sign for a definite plurality, without mediation—without translating to the word "two." A similar intention is involved in recognizing

𝍬

as a definite plurality, as a gestalt, without translating to "five."

Now I ask you to apply this sort of hermeneutic to Necker-cube stroke-numerals. I ask you to grasp the sign-row as a numeral, as a gestalt. (Without using ordinary counting to call off the strokes.) For a two-place display, you are to take such images as

and

as plurality-names without translating into English words. (Similarly for

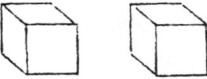

in the case where I choose to read "blank" as "zero.") Perhaps it is necessary to spend considerable time with this new symbolism before

---

*Because this notion corresponds to a situation in which we are unable to appraise image-rows as numerals, as gestalts.

recognition is achieved. Again, I encourage the reader to make short displays for practice. I have set a display of nine figures as the upper limit for which it might be possible to learn to grasp every sign-row as a numeral, as a gestalt.

The circumstance that the apprehended numeral may be different the next moment is not a mistake; the apprehended numeral is supposed to be in flux. So when you see image-rows, you take them as identical signs/semantics for the appearing pluralities.

But who wants such numerals—where are there any phenomena for them to count? For one thing, they count the very image-rows which constitute them. The realm of these image-rows is a realm of subjective flux: its plurality is authentically represented by my numerals, and cannot be authentically represented by traditional arithmetic.

A further remark which may be helpful is that here numerals arise only visually. So far, my numerals have no phonic or audio equivalent. (Whereas Freudenthal in effect posited an audio version of Hilbert numerals, using beeps.)

To repeat, by the "ostensive hermeneutic" I mean grasping the sign-row, without mediation, as a numeral. But there is, as well, the point that the Necker-cube numerals are *ostensive numerals*. That is, the (momentary) numeral for six would in fact be an image-row with just six occurrences of the image "upward cube." (Compare e.g. III⊃⊂IICI) The numeral is a collection in which only the "copies" of "upward cube" contribute positively, so to speak; and these copies demand a count of six (bicuturally). This feature needs to be clear, because later I will introduce numerals for which it does not hold.

Let me add another proviso concerning the ostensive hermeneutic which will be important later. I will illustrate the feature in question with an example which, however, is only an analogy. Referring to Arabic decimal-positional numerals, you can appraise the number-name of

$$1001$$

(comma omitted) immediately. But consider

$$786493015201483492147$$

Here you cannot appraise the number-name without mediation. That is, if you are asked to read the number aloud, you don't know whether to begin with "seven" or "seventy-eight" or "seven hundred eighty-six."

Lacking commas, you have to group this expression from the right, in triples, to find what to call it. An act of analysis is required.

In the case of Necker-cube numerals and the ostensive hermeneutic, I don't want you to see traditional number-names in the pluralities. However, I ask you to grasp a sign-row as a numeral, as a gestalt. I now add that the gestalt appraisal is definitive. I rule out appraising image-rows analytically (by procedures analogous to mentally grouping an Arabic number in triples). (I established a display of nine figures as the upper limit to support this.)

The need for this proviso will be obscure now. It prepares for a later device in which, even for short displays, gestalt appraisal and appraisal by analysis give different answers, either of which could be made binding.

* * *

The bicultural hermeneutic is applied, in effect, in my uninterpreted calculus "Derivation," which serves as a simplified analogue of my early concept art piece "Illusions." (Refer to the reproductions on the next four pages.) Strictly, though, "Derivation" does not concern a Necker-cube stroke-numeral. The individual figures are not Necker cubes, but "Wedberg cubes," formed with some foreshortening to make one of the two orientations more likely to be seen than the other. What is of interest is not apprehension of image-rows as numerals, but rather appraisal of lengths of the image-rows via ordinary counting. As for the lessons of this piece, a few simple observations are made in the piece's instructions. But to pursue the topic of concept art as uninterpreted calculi, and derive substantial lessons from it, will require an entire further study—taking off from earlier writings on post-formalism and uncanny calculi, and from my current writings collateral to this essay.

1987 Concept Art — Henry Flynt

"DERIVATION" (August 1987 corrected version)

Purpose: To provide a simplified analogue of my 1961 concept art piece "Illusions" which is discrete and non-"warping."* Thereby certain features of "Illusions" become more clearly discernible.

Given a perceptually multistable figure, the "Wedberg cube," which can be seen in two orientations: as a cube; as a prism (trapezohedron.)
Call what is seen at an instant an *image*.
Nine figures are concatenated to form the *display*.

An *element* is an image of the display for as long as that image remains constant (Thus, elements include: the image from the first instant of a viewing until the image first changes; an image for the duration between two changes; the image from the last change you see in a viewing until the end of the viewing.)

The *length* of an element equals the number of prisms seen. Lengths from 0 through nine are possible. Two different elements can have the same length. Length of element X is written $l(X)$.

Elements are seen in *temporal order* in the lived time of the spectator. I refer to this order by words with prefix 'T'. T-first; T-next; etc.

Element Y *succeeds* element X if and only if
i) $l(X) = l(Y)$, and Y is T-next after X of all elements with this length; *or*
ii) Y is the T-earliest element you ever see with length $l(X) + 1$.
Note that (ii) permits Y to be T-earlier than X: the relationship is rather artificial.

The *initial element* A is the T-first element. ($l(A)$ may be greater than 0; but it is likely to be 0 because the figure is biased.)

The *conclusion* C is the T-earliest element of length 9 (exclusive of A in the unlikely case in which $l(A) = 9$).

A derivation is a series of elements in lived time which contains A and C and in which every element but A succeeds some other element.

Discussion
To believe that you have seen a derivation, you need to keep track that you see each possible length, and to force yourself to see lengths which do not occur spontaneously.

You may know that you have seen a derivation, without being able to identify in memory the particular successions.

"Derivation" is not isomorphic to "Illusions" for a number of reasons. "Illusions" doesn't require you to see individually every possible ratio between the T-first ratio and unity. "Illusions" allows an element to succeed itself. The version of "Derivation" presented here is a compromise between mimicking "Illusions" and avoiding a trivial or cluttered structure. Any change such as allowing elements to succeed themselves would require several definitions to be modified accordingly.

---

*In "Illusions," psychic coercion, which may be called "false seeing" or "warping," is recommended to make yourself see the ration as unity. In "Derivation," this warping is not necessary; all that may be needed is that you see certain lengths willfully.

"DERIVATION"

THE DISPLAY

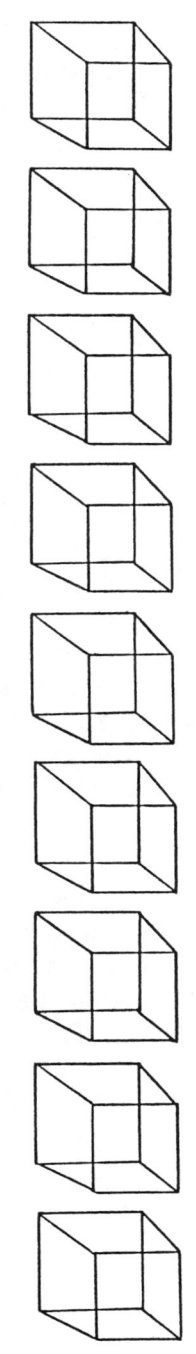

Concept Art Version of Mathematics System 3/26/61(6/19/61)

An "element" is the facing page (with the figure on it) so long as the apparent, perceived, ratio of the length of the vertical line to that of the horizontal line (the element's "associated ratio") does not change.

A "selection sequence" is a sequence of elements of which the first is the one having the greatest associated ratio, and each of the others has the associated ratio next smaller than that of the preceding one. (To decrease the ratio, come to see the vertical line as shorter, relative to the horizontal line, one might try measuring the lines with a ruler to convince oneself that the vertical one is not longer than the other, and then trying to see the lines as equal in length; constructing similar figures with a variety of real (measured) ratios and practicing judging these ratios; and so forth.) [Observe that the order of elements in a selection sequence may not be the order in which one sees them.]

An elaboration of "Stroke-Numeral" should be mentioned here, the piece called "an Impossible Constancy." (Refer to the facing page.) As written, this piece presupposes the bicultural hermeneutic, and that is probably the way it should be formulated. The point of this piece, paradoxically, is that one seeks to annul the flux designed into the apprehended numeral. Viewing of the Necker-cube numeral is placed in the context of a lived experience which is interconfirmationally weak: namely, memory of past moments within a dream (a single dream). Presumably, appraisals of the numeral at different times could come out the same because evidence to the contrary does not survive. So inconstancy passes as constancy. Either hermeneutic can be employed; but when I explained the hermetic hermeneutic, I encouraged you to follow the flux. Here you wouldn't do that—you wouldn't stare at the display over a retentional interval.

As for the concept of *equality* with regard to Necker-cube numerals, what can be said about it at this point? We have equality of numbers of figures in displays, by ordinary counting. We have two hermeneutics for identifying an apprehended numeral. In the course of expounding them, I expounded equivalence of different permutations of "stroke" and "vacant." Nevertheless, given that, for example, a display of two figures can momentarily count the numeral apprehended from a display of three figures,* we are in unexplored territory. Cross-tallying, suitable for judging equality of Hilbert numerals, seems maladapted to Necker-cube numerals; in fact, I dismissed it when introducing the ostensive hermeneutic.

If the "impossible constancy" from the paragraph before last were manageable, then one might consider restricting the ultimate definition of equality to impossible constancies. That is, with respect to a single display, if one wanted to investigate the intention of constancy (self-equivalence of the apprehended numeral), one might start with the impossible constancy. Appraisals of a given display become constant (the numeral becomes self-equivalent) in the dream. Then two *displays* which are *copies* might become constantly equivalent to each other, in the dream.

Such is a possibility. To elaborate the basics and give an incisive notion of equality is really an open problem, though. Other avenues might require additional devices such as the use of figures with distinctions of appearance.

---

*that it is not assured that copies of a numeral will be apprehended or appraised correlatively

1987 Concept Art — Henry Flynt

Necker-Cube Stroke-Numeral: AN IMPOSSIBLE CONSTANCY

The purpose of this treatment is to say how a Necker-cube stroke numeral may be judged (from the standpoint of private subjectivity) to have the same value at different times; even though the conventional belief-system says that the value is likely to change frequently.

This is accomplished by selecting a juncture in an available mode of illusion, namely dreaming, which annuls any distinction between an objective circumstance, and the circumstance which exists according to your subjective judgment. In the first instance, I don't ask you to change your epistemology. Instead, to repeat, I select an available juncture in lived experience at which the conventional epistomology gets collapsed.

You have to occupy yourself with the stroke-numeral to the point that you induce yourself to dream about it.

When, in apprehending a stroke-numeral, you "judge" the value of the numeral, the number, this refers to the image you see and to the number-word which you may conclude from the image.

Suppose that in a single dreamed episode, you judge the value of the numeral at two different moments. Suppose that at the second moment, you do not register any discrepancy between the value at the second moment and what the value was at the first moment. Then you are permitted to disregard fallibility of memory, and to conclude that the values were the same at both moments: because if your memory has changed the past, it has done so tracelessly. A tracelessly-altered past may be accepted as the genuine past.

Refinements. The foregoing dream-construct may be "lifted" to waking experience, as per the lengthy explanations in "An Epistemic Calculus." Now you are asked to alter your epistemology, selectively to suspend a norm of realism.

Now that we are concerned with waking experience, a supporting refinement is possible. Suppose I make an expectation (which may be unverbalized) that the value of the numeral at a future moment will be the same that it is now. This expectation cannot be proved false, if: the undetermined time-reference "future moment" is applied only at those later moments when the value is the same as at the moment the expectation was made. (Any later moment when the value is not the same is set aside as not pertinent, or forgotten at still later moments when the value is the same.)

As a postscript, there is another respect in which testing a fact requires trust in a comparable fact. Suppose I make a verbalized expectation that the value of the numeral in the future will be the same as at present. Then to test this expectation in the future depends on my memory of my verbalization. My expectation cannot be belied unless I have a sound memory that the number I verbalized in my expectation is different from the number I conclude from the image now.

III. Inconsistently-Valued Numerals

As the "Wedberg cube" illustrates, a cubical frame can be formed in different ways, altering the likelihood that one or another image is seen. With respect to the initial uses of the Necker-cube stroke-numeral a figure is wanted which lends itself to the image of a cube projecting up, or of a cube projecting down, with an approximately equal likelihood for the two images—and which makes other images unlikely. Now let a Necker cube be drawn large, with heavy line-segments, with all segments equally long, with rhomboid front and back faces; and display it below eye level.

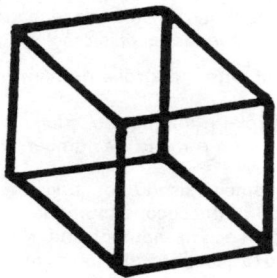

As you look for the up and down orientations, there should be moments when paradoxically you see the figure taking on both of these mutually-exclusive orientations at once—yielding an apparition which is a logical/geometric impossibility. The sense-content in this case is dizzying.

That we have perceptions of the logically impossible when we suffer illusions has been mentioned by academic authors. (Negative afterimages of motion—the waterfall illusion.) Evidently, though, these phenomena are so distasteful to sciences which are still firmly Aristotelian that the relations of perception, habituation, language, and logic manifested in these phenomena have never been assessed academically. For me to treat the paradoxical image thoroughly here would be too much of a digression from our subject, the apprehension of plurality. However, a sketchy treatment of the features of the impossible image is necessary here.

To begin with, the paradoxical image of the Necker cube is not the same phenomenon as the "impossible figures" shown in visual perception textbooks. The latter figures employ "puns" in perspective coding such that parts of a figure are unambiguous, but the entire figure

The Apprehension of Plurality

cannot be grasped as a gestalt coherently. Then, the paradoxical Necker-cube image is not an inconsistently oriented *object* (as the reader may have noted). It is an *apparitional depiction* of an inconsistently oriented object. But this is itself remarkable. For since a dually-oriented cube (in Euclidean 3-space) is self-contradictory by geometric standards, a picture of it amounts to a non-vacuous semantics for an inconsistency. Another way of saying the same thing is that the paradoxically-oriented image is real *as an apparition*.

If one is serious about wanting a "logic of contradictions"—a logic which admits inconsistencies, without a void semantics and without entailing everything—then one will not attempt to get it by a contorted weakening of received academic logic. One will start from a concrete phenomenon which demands a logic of contradictions for its authentic representation—and will let the contours of the phenomenon shape the logic.

In this connection, the paradoxically-oriented Necker-cube image provides a lesson which I must explain here. Consider states or properties which are mutually exclusive, such as "married" and "bachelor." Their conjunction—in English, the compound noun "married bachelor"—is inconsistent.* On the other hand, the joint denial "unmarried nonbachelor" is perfectly consistent and is satisfied by nonpersons: a table is an unmarried nonbachelor. "Married" and "bachelor" are mutually exclusive, but not exhaustive, properties. Only when the domain of possibility, or intensional domain, is *restricted* to persons, so "married" and "bachelor" become exhaustive properties.** Then, by classical logic, "married bachelor" and "unmarried nonbachelor" both have the same semantics: they are both inconsistent, and thus vacuous, and thus indistinguishable. For exhaustive opposites, joint affirmation and joint denial are identically vacuous.

But the paradoxically-oriented Necker-cube image provides a concrete phenomenon which combines mutually exclusive states—as an apparition. We can ascertain whether a concrete case behaves as the tenets of logic prescribe. As I have said, various images can be seen in a Necker cube, including a flat image. Thus, the "up" and "down" cubes

---

*If I must show that it is academically permitted to posit notions such as these, then let me mention that Jan Mycielski calls "triangular circle" inconsistent in *The Journal of Symbolic logic,* Vol. 46, p. 625.

**I invoke this device so that I may proceed to the main point quickly. If it is felt to be too artificial, perhaps it can be eliminated later.

are analogous to "married" and "bachelor" in that they are not exhaustive of a domain unless the domain is produced by restriction. Then "neither up nor down" is made inconsistent. (It is very helpful if you haven't learned to see any stable images other than "up" and "down.") The great lesson here is that given "both up and down" and "neither up nor down" as inconsistent, their concrete reference is quite different. To see a cube which manifests both orientations at the same time is one paradoxical condition, which we know how to realize. To see a cube which has no orientation (absence of "stroke" and absence of "vacant" both) would be a different paradoxical condition, which we do not know how to realize and which may not be realizable from the Necker-cube figure. I don't claim that this is fully worked out; but it intimates a violation of classical logic so important that I had to mention it. *When concept art reaches the level of reconstructing our inferential intuitions as well as our quantitative intuitions, such anomalies as these will surely be important.*

Referring back to the Necker cube of page 210, let us now intend it as a stroke-numeral (display of one figure). Let me modify the previous assignments and stipulate that "blank" means "zero," rather than "no numeral present." (It is more convenient if every sign yields a numeral.) When you see the paradoxical image, you are genuinely seeing "a" numeral which is the simultaneous presence of two mutually exclusive numerals "one" and "zero"—because it is the simultaneous presence of images which are mutually exclusive geometrically.***

It's not the same thing as

$$1$$

—or as an alternative,

$$0$$

—because these are merely ambiguous scripts. In the Necker-cube case, two determinate images which by logic preclude each other are present at once; and as these images are different numerals, we have a genuine

---

*For brevity, I may compress the three levels image, sign, numeral in exposition.

The Apprehension of Plurality

inconsistently-valued numeral.

This situation changes features of the Necker-cube numerals in important ways, however. Lessons from above become crucial. We transfer the ostensive hermeneutic to the new situation, and find an inconsistent-valued numeral. But this is no longer an *ostensive numeral*. We have a name which is one and zero simultaneously, but this is because of the impossible shape (orientation) of the notation-token. What we do not have is a collection of images of a single kind (the stroke) which paradoxically requires a count of one and a count of zero. "Stroke" is positively present, while "vacant" is positively present in the same place. We will find that a display with two figures can be inconsistent as zero and two; but it is not an ostensive numeral, because the number of strokes present is two uniquely.* Here the numerals are not identically their semantics: for the anomaly is not an anomaly of counting. The ambiguous script numeral is a proper analogy in this respect. To give an anomaly of counting which serves as a concrete semantics for the inconsistently-valued numerals, I will turn to an entirely different modality.

From work with the paradoxical image, we learn that the Necker cube allows some apprehensions which are not as common as others—but which can be fostered by the way the figure is made and by indicating what is to be seen. These rare apprehensions then become intersubjectively determinate. If one observes Necker-cube displays for a long time, one may well observe subtle, transient effects. For example, you might see the "up" and "down" orientations at the same time, but see one as dominating the other. In fact, there are too many such effects and their interpersonal replicability is dubious. If we accepted such effects as determining numerals, the interpersonal replicability of the symbols would be eroded. Also the concrete definiteness of my anomalous, paradoxical effects would be eroded. So I must stipulate that every subtle transient effect which I do not acknowledge explicitly is not definitive, and is unwanted, when the display is intended as a symbolism.

Let me continue the explanation, for the inconsistently-valued

---

*Referring to my "person-world analysis" and to the dichotomy of Paradigm 1 and Paradigm 2 expounded in "Personhood III," this token which is two mutually exclusive numerals because its shape is inconsistent is outside that dichotomy: because established signs acquire a complication which is more or less self-explanatory, but the meanings do not follow suit.

numerals, for displays of more than one figure. When the display consists of two Necker cubes, and the paradoxical images are admitted, what are the variations? In the first place, one figure might be seen (in a moment) as a paradoxical image and the other as a unary image. Actually, if it is important to obtain this variant, we can compel it, by drawing one of the cubes in a way which hampers the double image. (Thin lines, square front and back faces, the four side segments much shorter than the front and back segments.) Then we stipulate that the differently-formed cubes continue to have the same assigned interpretation.

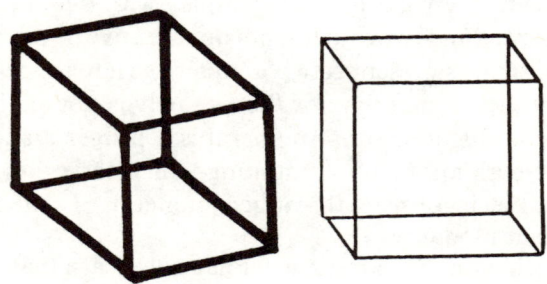

Reading the two-figure display, then, the paradoxical and unary images concatenate so that the resulting numeral is in one case one and two at the same time; and in the other case zero and one at the same time. Of course, it is only in a moment that either of these two cases will be realized. At other moments, one may have only unary images, so that the numeral is noncontradictorily zero, one, or two as the case may be. (If it is important to know that we can obtain a numeral which is both one and two at the same time without using dissimilar figures, then, of course, we can use a single figure and redefine the signs as "one" and "two.")

Now let us consider a display of two copies of the cube which lends itself to the paradoxical image. Suppose that two paradoxical images are seen; what is the numeral? Here is where I need the proviso which I introduced earlier. Every sign-row is capable of being grasped as a numeral, as a gestalt; and the appraisal of image-rows as numerals, analytically, is ruled out. Let me explain how this proviso applies when two paradoxical images are seen.

Indeed, let me begin with the case of a pair of ambiguous script-numerals:

11

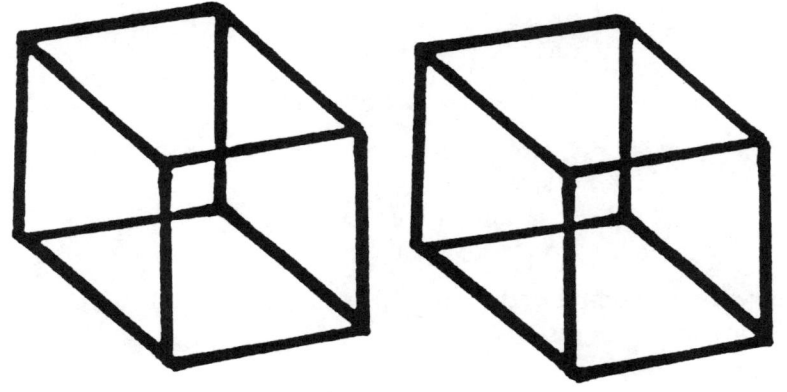

When these numerals are formed as exact copies, and I appraise the expression as a numeral, as a gestalt, then I see 11 or I see 22. ("Concatenating in parallel") I do not see 21 or 12—although these variants are possible to an analytical appraisal of the expression. In the gestalt, it is unlikely to intend the left and right figures differently. This case is helpful heuristically, because it provides a situation in which the perceptual modification is only a matter of emphasis (as opposed to imputation of depth). To this degree, the juncture at issue is externalized; and it is easier to argue a particular outcome. On the other hand, the mechanics differ essentially in the script case and the Necker-cube case.

In the Necker-cube case, one sees both the left and the right image determinately both ways at once. This case may be represented as

$$\left\{ \begin{array}{l} \text{stroke} \\ \text{vacant} \end{array} \right. \quad \left\{ \begin{array}{l} \text{stroke} \\ \text{vacant} \end{array} \right.$$

Analytically, then, four variants are available here,

> stroke-stroke
> stroke-vacant
> vacant-stroke
> vacant-vacant

However, to complete the present explanation, only two of these variants appear as gestalts,

> stroke-stroke
> vacant-vacant

I chose to rule out the three-valued numeral which would be obtained by analytically inventorying the permutations of the signs afforded in the perception. The two-valued numeral arising when the sign-row is grasped as a gestalt is definitive.

Let me summarize informally what I have established. Relative to a two-figure display with paradoxical images admitted, we have a numeral which is inconsistenly two and zero. We can also have a numeral which is inconsistently one and zero, and a numeral which is inconsistently two and one. (In fact, these variants occur in several ways.) But we don't have a numeral which is inconsistently zero, one, and two—even though such a variant is available in an analytical appraisal—because such a numeral does not appear, in perception, as a gestalt.

Academic logic would never imagine that there is a situation which demands just this configuration as its representation. Certain definite positive inconsistencies are available in perception. Other definite positive inconsistencies, very near to them, are not available. Once again, *if one wants a vital "logic of contradictions," one has to develop it as a representation of concrete phenomena; not as an unmotivated contortion of received academic logics.*

* * *

But what is the use of inconsistently-valued numerals? I shall now provide the promised concrete semantics for them. This semantics utilizes another experience of a logical impossibility in perception. This time the sensory modality is touch; and the experienced contradiction is one of enumeration. Aristotle's illusion is well known in which a rod, placed between the tips of crossed fingers, is felt as two rods. (Actually, the greater oddity is that when the rod is held between uncrossed fingers, it is felt as one even though it makes two contacts with the hand.) I now replace the rod with a finger of the other hand: the same finger is felt as one finger in one hand, as two fingers by the other hand. So the same entity is apprehended as being of different pluralities, in one sensory modality.

Let me introduce some notation to make it easier to elaborate. Abbreviate "left-hand" as L and "right-hand" as R. Denote the first, middle, ring, and little fingers, respectively, as 1, 2, 3, and 4. Now cross L2 and L3, and touch R3 between the tips of L2 and L3. One feels R3 as one finger in the right hand, and as two fingers with the left hand. As apparition, R3 gets a count of both one and two, apprehended in the same sensory modality at the same time. Here is a phenomenon authentically signified by a Necker-cube numeral which is both "1" and "2."

The crossed-finger device is obviously unwieldy. The possibilities can, however, be enlarged somewhat, to make a further useful point. For example, touch L1 and R3, while touching crossed L2 and L3 with R4. Here we have a plurality, concatenated from one unary and one paradoxical constituent, which numbers two and three at the same time.

Then, we may cross L1 and L2 and touch R3, while crossing L3 and L4 and touching R4. Now we have a plurality which is two and four at the same time. In terms of perceptual structure, it is analogous to the numeral concatenated from two paradoxical images. As gestalt, we concatenate in parallel. In the case of the fingers, we do not find a plurality of three unless we appraise the perception analytically (blocking concatenation in parallel).

If one wants the inconsistently-valued numerals to be ostensive numerals, then one can use finger-apparitions to constitute stroke-numerals. Referring back to the first example, if we specify that the stroke(s) is your R3-perception, or the apparition R3, then we obtain a stroke which is single and double at the same time. Now the inconsistently-valued numeral is identically its semantics: it authentically names the token-plurality which constitutes it.

I choose not to rely heavily on this device because it is so unwieldy. The visual device is superior in that considerably longer constellations are in the grasp of one person. Of course, if one chose to define fingers as the tokens of ordinary counting, one might keep track of numbers larger than ten by calling upon more than one person. The analogous device could be posited with respect to the inconsistently-valued numbers; but then postulates about intersubjectivity would have to be stated formally. I do not wish to pursue this approach.

It is worth mentioning that if you hold a rod vertically in the near center of your visual field, hold a mirror beyond it, and focus your gaze on the rod, then you will see the rod reflected double in the mirror. This is probably not an inconsistent perception, because the inconsistent counts don't apply to the same apparition. (But if we add Kant's postulate that a reflection exactly copies spacial relations among parts of the object, then the illusion does bring us close to inconsistency.) The illusion illustrates, though, that there is a rich domain of phenomena which support mutable and inconsistent enumeration.

IV. Magnitude Arithmatic

I will end this stage of the work with an entirely different approach to subjectively variable numerals and quantities. I use the horizontal-vertical illusion, the same that appeared in "Ilusions," to form numerals. The numeral called "one" is now the standard horizontal-vertical illusion with a measured ratio of one between the segments. The numeral called "two" becomes a horizontal-vertical figure such that the vertical has a measured ratio of two to the horizontal segment. Etc. If "zero" is wanted, it consists of the horizontal segment only.

The meaning of each numeral is defined as the apparent, perceived length-ratio of the vertical to the horizontal segment. Thus, for example, the meaning of the numeral called "one" admits subjective variation above the measured magnitude. For brevity, I call this approach magnitude arithmetic—although the important thing is how the magnitudes are realized.

The Apprehension of Plurality

In all of the work with stroke-numerals, numbers were determinations of plurality. An ostensive numeral was a numeral formed from a quantity of simple tokens, which quantity was named by the expression. The issue in perception was the ability to make gestalt judgments of assemblies of copies of a simple token.

The magnitude numerals establish a different situation. Magnitude numerals pertain to quantity as magnitude. They relate to plurality only in the sense that in fact, measured vertical segments are integer multiples of a unit length; and e.g. the apprehended meaning of "two" will be a magnitude always between the apprehended meanings of "one" and "three"—etc.

Once again we can distinguish a bicultural and an ostensive hermeneutic. The bicultural hermeneutic involves judging meanings of the numerals with estimates in terms of the conventional assignment of fractions to lengths (as on a ruler). I find, for example, that the magnitude numeral "two" may have a meaning which is almost 3. (Larger numerals become completely unwieldly, of course. The point of the device is to establish a principle, and I'm not required to provide for large numerals.)

Then there must be an ostensive hermeneutic, a "magnitude-ostensive" hermeneutic. Here the subjective variations of magnitude do not receive number-names. They are apprehended (and retentionally remembered) ostensively.

As I pointed out, above, the concept of equality with regard to Necker-cube numerals is at present an open problem. To write an equality between two Necker-cube displays of the same length is not obviously cogent; in fat, it is distinctly implausible. For magnitude numerals, however, it is entirely plausible to set numbers equal to themselves—e.g.

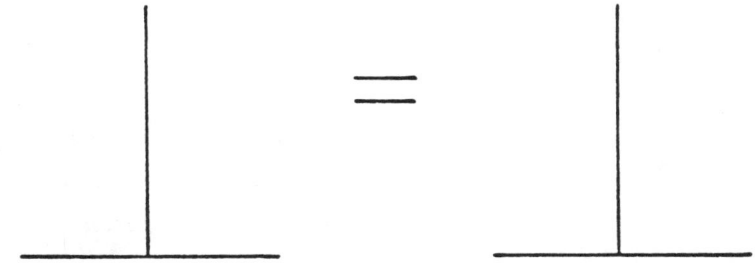

The point is that it is highly likely that copies of a magnitude numeral will be apprehended or appraised correlatively. This was by no means guaranteed for copies of a Necker-cube numeral displayed in proximity.

$$\vdash + \vdash = \vdash$$

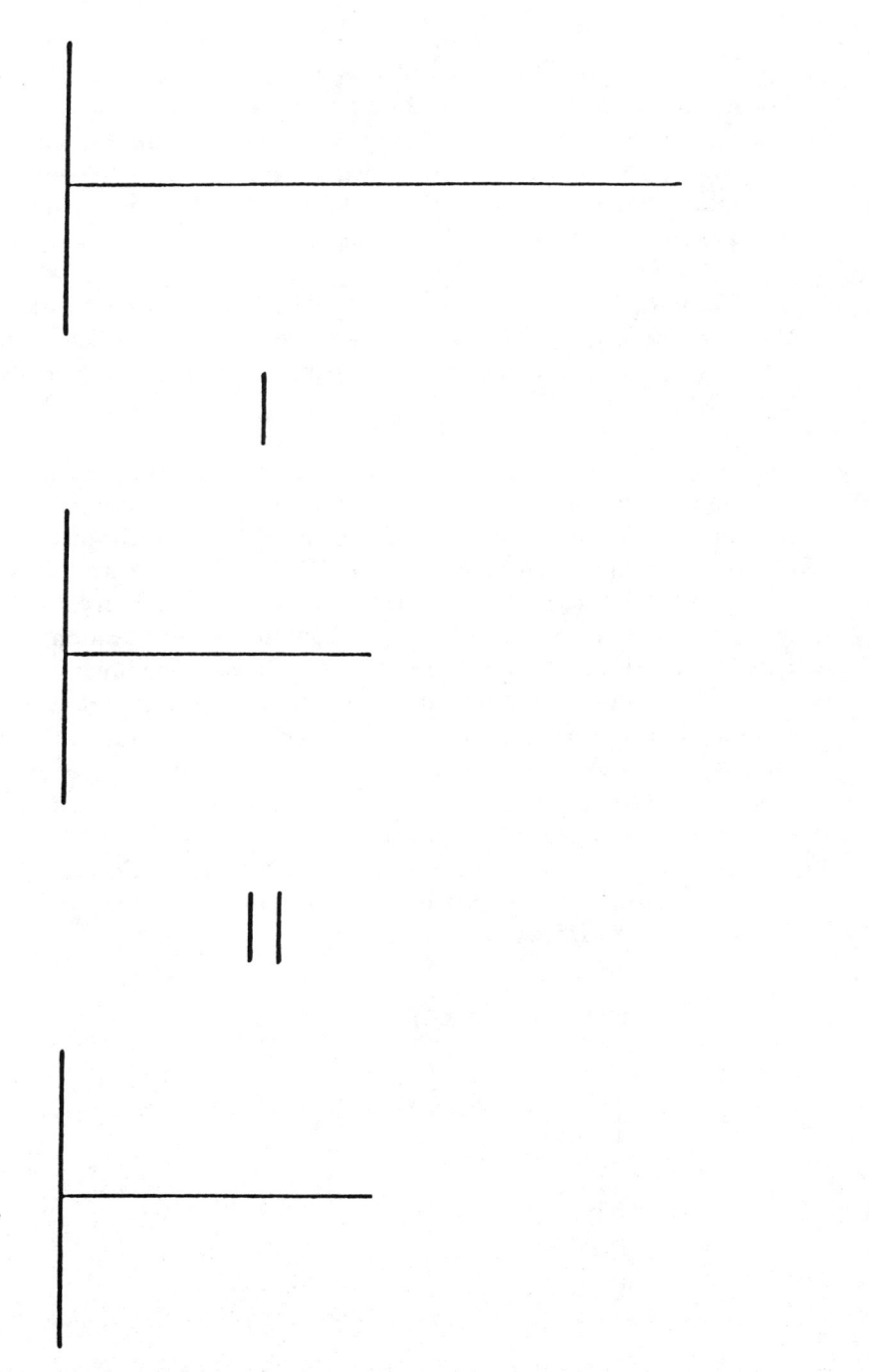

Upon being convinced that these simplest of equations are meaningful, we may stipulate a simple addition, "one" plus "one" equals "two." (It was not possible to do anything this straightforward with Necker-cube numerals.) Continuing, we may write a subtraction with these numerals. There may now appear a complication in the rationale of combination of these quantities. The "two" in the subtraction may appear shorter than the "two" in the addition. A dependence of perceptions of these numbers on context may be involved.

We find, further, that "readings" of these equations according to the bicutural hermeneutic yield propositions which are false when referred back to school-arithmetic—e.g. the addition might be read as

$$1^1/_5 + 1^1/_5 = 2^4/_5$$

So the effect of inventing a context in which a relationship called "one plus one equals two" is appraised as $1^1/_5 + 1^1/_5 = 2^4/_5$ (where there is a palpable motivation for doing this) is to erode school-arithmetic.

Another approach to the same problem is to ask whether magnitude arithmetic authentically describes any palpable phenomenon. The answer is that it does, but that the phenomenon in question is the illusion, or rationale of the illusion. The significant phenomenon arises from having both a measured ratio and a visually-apparent ratio, which diverge. This is very different from claiming equations among non-integral magnitudes without any motivation for doing so. Indeed, given that the divergence is the phenomenon, the numerals are not really ostensive in a straightforward way.

One way of illustrating the power of the phenomenon which models magnitude arithmetic is to display ruler grids flush with the segments of a horizontal-vertical figure.

The Apprehension of Plurality

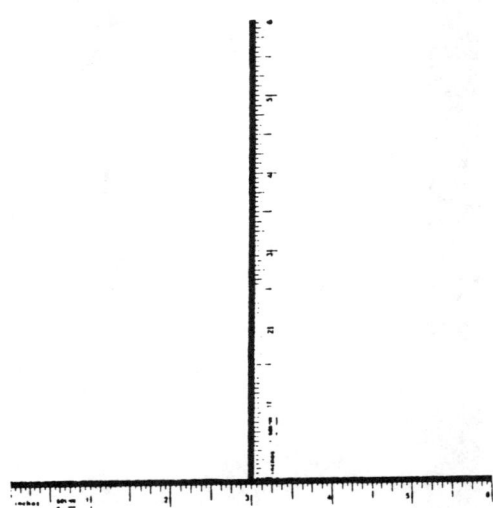

What we find is that the illusion visually captures the ruler grids: it withstands objective measurement and overcomes it. We have a non-trivial, systematic divergence between two overlapping modalities for appraising length-ratios—one modality being considered by this culture to be subjective, and the other not.

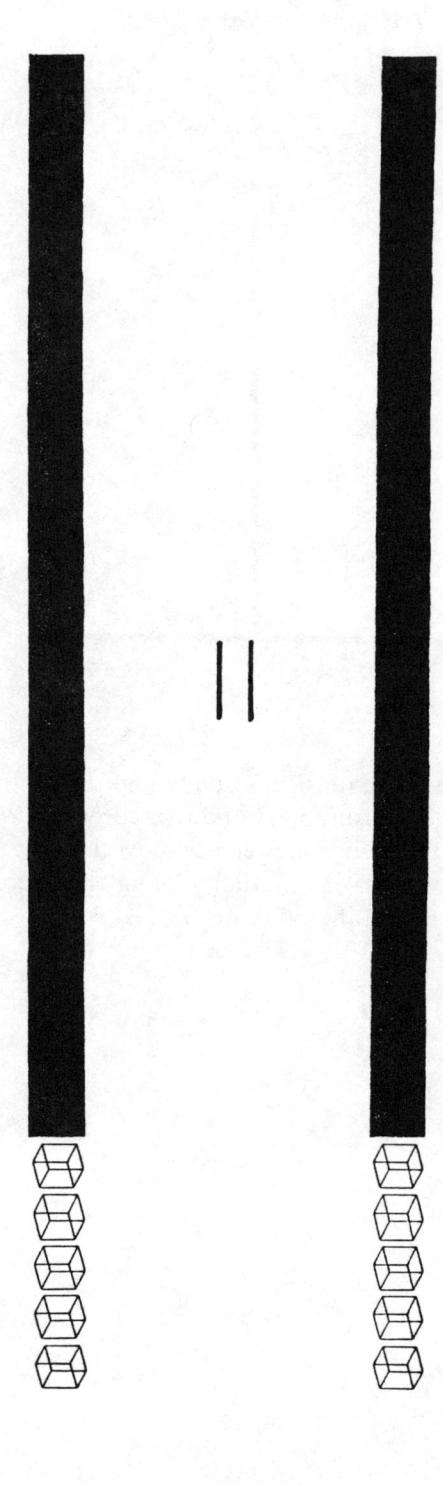

The Apprehension of Plurality

In "Derivation" I used multistable cube figures to give a simplified, discrete analogue of the potentially continuous "vocabulary" in "Illusions." I could try something similar for magnitude numerals. Take as the magnitude unit a black bar representing an objective unit of twenty 20ths, concatenated with a row of five Necker cubes. Each cube seen in the "up" orientation adds another 20th to the judged magnitude of the subjective unit, so that the unit's subjective magnitude can range to $1\frac{1}{4}$. When, however, we write the basic equality between units, it becomes clear that this device does not function as it is meant to. In particular, the claim of equality applied to the Necker-cube tails is not plausible, because it is not guaranteed that these tails will be apprehended or appraised correlatively. I have included this case as another illutration of the sort of inventiveness which this work requires; and also to illustrate how a device may be inadequate.

*   *   *

This completes the present stage of the work. Let me emphasize that this manual does little more than define certain devices developed in the summer of 1987. These devices can surely give rise to substantial lessons and substantial applications.

There is my pending project in *a priori* neurocybernetics. Given that mechanistic neurophysiology arrives at a mind-reading machine—called, in neurophysiological theory, an autocerebroscope—devise a text for the human subject such that reading it will place the machine in an impossible state (or short-circuit it). Such a problem is treated facetiously in Raymond Smullyan's *5000 B.C.;* and more seriously by Gordon G. Globus' "Mind, Structure, and Contradiction," in *Consciousness and the Brain,* ed. Gordon Globus *et al.* (New York, 1976), p. 283 in particular. But I imagine that my Necker-cube notations will be the key to the first profound, extra-cultural solution.

In any case, this essay is only the beginning of an enterprise which requires collateral studies and persistence far into the future to be fulfilled. (I may say that I first envisioned the possibility of the present results about twenty-five years ago.)

Background References

David Hilbert, three papers in *From Frege to Gödel,* ed. Jean van Heijenoort (1967)
David Hilbert, "Neubegrundung der Mathematik" (1922)
David Hilbert and P. Bernays, *Grundlagen der Mathematik* I (Berlin, 1968), pp. 20–25
Plato, "Philebus"
Aristotle, *Metaphysics,* I.6
Proclus, *A Commentary on the First Book of Euclid's Elements,* tr. Glenn Morrow (Princeton, 1970), 54–55
Hans Freudenthal, *Lincos: Design of a Language for Cosmic Intercourse* (Amsterdam, 1960), pp. 14–5, 17, 21, 45–6
Kurt Gödel in *The Philosophy of Bertrand Russell,* ed. Paul Schilpp (1944), p. 137
W.V.O. Quine, *Mathematical Logic* (revised), pp. 121–2
Paul Benacerraf, "What numbers could not be," in *Philosophy of Mathematics* (2nd edition), ed. Paul Beneacerraf and Hilary Putnam (1983)
Leslie A. White, "The Locus of Mathematical Reality: An Anthropological Footnote," in *The World of Mathematics,* ed. J.R. Newman, Vol. 4, pp. 2348–2364
Herman Weyl, *Philosophy of Mathematics and Natural Science* (Princeton, 1949), pp. 34–7, 55–66
Andrei Markov, *Theory of Algorithms* (Jerusalem, 1961)
G.T. Kneebone, *Mathematical Logic and the Foundations of Mathematics* (London, 1963), p. 204ff.
Michael Resnik, *Frege and the Philosophy of Mathematics* (Ithaca, 1980), pp. 82, 99
Ludwig Wittgenstein, *Wittgenstein's Lectures on the Foundations of Mathematics* (1976), p. 24; but p. 273
Ludwig Wittgenstein, *Philosophical Grammer* (Oxford, 1974), pp. 330–331
Steven M. Rosen in *Physics and the Ultimate Significance of Time,* ed. David R. Griffin (1986), pp. 225–7
Edgar Rubin, "Visual Figures Apparently Incompatible with Geometry," *Acta Psychologica,* Vol. 7 (1950), pp. 365–87
E.T. Rasmussen, "On Perspectoid Distances," *Acta Pschologica,* Vol. II (1955), pp. 297–302
N.C.A. da Costa, "On the Theory of Inconsistent Formal Systems," *Notre Dame Journal of Formal Logic,* Vol. 15, pp. 497–510
F.G. Asenjo and J. Tamburino, "Logic of Antinomies," *Notre Dame Journal of Formal Logic,* Vol. 16, pp. 17–44

Richard Routley and R.K. Meyer, "Dialectical Logic, Classical Logic, and the Consistency of the World," *Studies in Soviet Thought,* Vol. 16, pp. 1–25

Nicolas Goodman, "The Logic of Contradiction," *Zeitschr. f. math. Logik und Grundlagen d. Math.,* Vol. 27, pp. 119–126

Hristo Smolenov, "Paraconsistency, Paracompleteness and Intentional Contradictions," in *Epistemology and Philosophy of Science* (1982)

J.B. Rosser and A.R. Turquette, *Many-valued Logics* (1952), pp. 1–9

Gordon G. Globus, "Mind, Structure, and Contradiction," in *Conciousness and the Brain,* ed. Gordon Globus *et al.* (New York, 1976), p. 283

Poems

George Quasha

REFERENCE IS THE ONE IMPOSSIBLE THING

> To name it
> makes it
> what in it
> is it

George Quasha

MIXING PERSONS

When you
said
what you
said
to me
to your-
self
you con-
fused
us

ON THE ALERT

to catch
it quick
before
even
it
can be
said
to be
thought

George Quasha

MORPHIC RESONANCE

Form
sounds
name
like
sound
forms
shape
things

THIS SENTENCE IS NOT SO LONG

as
to
make
it-
self
so
de-
vious-
ly ex-
tended
you
lose
track
at
any
point
here-
with
anti-
cipated
where
you
de-
cide
to
stop
and look
back
to
where
you
were
just
now
and
can-
not

George Quasha

THIS ISN'T LANGUAGE

    my
    foot

Poems

HERMENEUTICALLY SPEAKING

In the beginning
is
not
how
it read
the first time

George Quasha

ALLONYM

The
name
the
other
one
you
are
knows
you
by

THE ALLOMORPHIC SET =

[the
one
thing
the
more
than
one
thing
have
in
common
that
each
one
thing
has
differently]

George Quasha

SITTING

re-
flecting
up-
on
the
water

DON'T LOOK NOW

Nothing
is
back there.

**LOOKING
ON AN-
TERIOR I-
DENTITY**

Do
you
remember
remembering
remembering
the first
time

LOOKING MORE LIKE IT

The stranger
it
comes
into town
the more
it
looks
like
life it-
self

George Quasha

THE SLANT I TAKE AS STANCE

If it's
good
enough
to stand
on
it's good
enough
to let
loose
in
through

THE MAN WHO DISAPPEARED

let
the light
shine
clear
through

George Quasha

THOROUGH THROUGH & THROUGH

EVERYTHING HAS MANY VERSIONS

unnoticed here

George Quasha

EXTENDED SPEAKING

However many
times you say it
it makes
itself all
over

WHAT IS WHERE AND NOT

If creation
is out
of nothing,
nothing is
more
immanent
in the thing
than the thing is
immanent
in itself

George Quasha

IT'S ONLY HOPE

is
that it is
other
than itself

for other-
wise
it is not
what it is

THE WORD IS

the
world
erasing
its
name
before
your
eyes
register
the critical
letter
lacking

George Quasha

(for Harry Smith)

THIS IS NOTHING

next to you.
Exposing portraits of people hiding in photo-
emulsion painted in
Signed the Eyer's Hand.
Sleepers
Awake.
NEXT AREA
34 MILES

I really wish what you told me
is true. Only any
meaning is repression. Power is a pain in the ass.
What's the difference?
Extended Low Grade Confusion (ELGC).

THE LIMITS ARE NOT
known. Now we meet the outer bounds of a state
ment for you (you on the inside):

If I could say something you could not
under
stand but accept anyway (break
down). I am next to you
nothing (how very particular

neither of us is in a position to know
but to accept
travels another level) and into the next area

that for no good reason comes here. Reading
against
the grain. Winning his way.
Double you sounds
a letter making a difference between two. MORPHIC

If I were to acknowledge a lifelong failure it would be
not knowing the difference
between two. You's. I's.
Is organs. Like. driving.
at. orthogonals. to. the. double. line.
Keeping straight on the middle. RESONANCE

ONE WAY ONLY
Seeing God *without* means in His own being.
Eckhart is unreasonable.
Automobile: moves itself.
Dreamt dream as mode of habitation.
Rhythmically alternating recurring altered states
(bounds) in Mr. Smith comes to heaven
(or earth) magic. Cellular celluloid.
History is telling
showing up.

Crossing
swords properly in contact never build up
more than four ounces.
Go.on.while.you.have.the.light. (John 12:35)
Speaking degree ought out of mind-degradable personality.
Situations are syntactic. In combat
a person is struck only by his own sword
returning home. Likewise the sounds of voices. No
one is in charge. Only specific weight.

Driving conditions hazardous. WEATHER REPORT:
Seasonably cruel. MAKING THINKING
think within the reflexivity of its own
means. Blades. Frames. Sit-
uations. She mis laid her hus band some where a
long the line. W i t h i n l i m i t s . moving
lights. SUDDEN TRANSITION: she started running her
self. NEVER TO KNOW HOW IT LOOKS BEFORE
it
looks.
SHIVA'S ARMED
forces.

George Quasha

Let's get out of here and go for a nice contextual walk.
With luck we'll get a god's eye view of this
blasted ego (look up
yourself [as in reference]). IT LOOKS BEFORE
it
leaps.
A poem is a kind of prayer to remain
readerless. GETTING SOMEWHERE: He rode
his life far
more than any other. Love happens in no time.

Everything said is keyed in each thing said. Sharp.
I underscored him on his performance. For life is up
in no time
flat. You can't move away as fast as it moves
in you. Who cares? You cares. Makes three of us.
If you knew what you know they'd never believe it.
Did the hand that writes know what it said?
No. You know. It could stop anywhere.
That it doesn't means
that it doesn't. Say something.
Ask me anything. Want to know who I am? Come
on in. Once there was
a young person who came to this city in this state
to tell a tale and the people were willing to listen and
this is what they said.
If you could move as fast as it moves in itself you'd be it
and you are but you can't. You speak kitchen latin?
Eat your words. *Durchbruch*. Utensil: an in.strument
of in.vestigation in.tending to in.form. Confession:
I let go to sink, the word went free, I was still alone
gurgling. O Sad One lend us this night our unread book.

Heaven is now not then after all. So one must negotiate
the curves so to speak circumferentially. Squinting
thinking. Living hand to nose
eating you smell but do you smelling also eat?
That proverbial dystax. An inflamation of the circum
stance. Mode of loco
motion. Instrum

ent. Following is inter
pretation. *Durch* (through to no end) *bruch*
(break to no beginning). The Master
is a thin man with a heart the size of the eye
of a needle upon which three times four angels
dance as we speak. We sit in silence
reflecting upon the water.
If I may call attention to this placed concreteness
(once I made it I knew it was). Disappearing
the name before your eyes. It is not what it is
otherwise. It makes itself move itself
all over (serpent in the garden). Called
speaking. And let loose through to thoroughly.
She spread her ideas for me. *Deja-lu*.
Good enough to lie in, remembering remembering.
Closet schizos refusing to call themselves
lose allonyms like socks in no time
like the present. You disappear
where you can't read.
The subjunctive sublimental performative I promise
to be here in the gap
next to you, not knowing
in the nothing in the light in the light
neither where it touches nor where it wells forth nor
where it rests in it
self

Poems from
The Sad World

Charles Stein

(*The Sad World* is a region from *theforestforthetrees*)

Poems from The Sad World

The Sad Trees Rows
(for Harry Smith)

*Our thoughts are tokens
of what we think...*

1

You see trees
grow on an endless road.

There is no end to the trees that grow there.
One in each place. One place

for every tree

that grows

on an endless road.

...

In a little bus

provided to survey

the endless rows

of trees

on an endless road—

    out riding.

Outriding the long long rows that never end.
Outriding the long long night—

Black.

Blank.

A bed of rose trees.

Charles Stein

Blank.
Black. A head of
rose trees
grows blank.
        Black
velvet back cloth
for display of a bed of rose
trees. Outriding the bed of night
to the end of rows.
. . .
At the end of night
the minds that hold the thoughts that know
confer
to end the night that holds them blind and blank.

The thought that only knows the place it holds
in a night of thought.

A mouth enclosing the night.
A thought inside a mouth.
. . .
The minds confer to end the night — to fold
themselves
up in the blank
blanket of place that holds them
up — *sus*tains and *re*strains —
pains them.

Poems from The Sad World

It is plain to them.

. . .

The minds confer to hold the place

about them up—

to *have* the thought that chains them,

sustains and constrains

them in and to

the tokens they have

that cannot hold

the blank night up.

. . .

All trees—

you cannot see

or count them

up

     pass

into the night of rows.

2

The words at the ends of the words trees watch the blanks

of space up-hold them.

The minds that cannot hold themselves

confer to watch the trees rows grow mind's words.

One of them leaves the rows—the blank

rank of mind's trees—

Charles Stein

Trees leaves fall on a bed of rows.

3

Were I myself I do not think to speak of it.

I would not come to the end of all night's rows.

Moving one at a time
or passing without bounds—

The trees rows fold
into the place that holds them
up in the tokens of thought.

They lose themselves in a bed of falling trees rows.
They fail to find themselves again beneath the leaves.

The autumn tokens fall on a bed of night's last rows.

Later Poems from *The Sad World*

9/24/87

Passing from one locality to another

and passing so by utilizing

some manner of conveyancing

                some process.

Being one among

an assortment of some such beings

and being transformed

           by some

        manner of transformation

    into some

                one   other

            being

      among

    an assortment

        of other such beings.

And coming back again

and being changed back again

traveling by reverse conveyance—
the car goes backwards
from the town at which
                 one
            had
         just recently
              established oneself
         back to that town
                  from
which
recently one
    had   departed.

Or turning back into the thing one used to be
    by the backward of the process that
        just recently
            had transformed one.

Or turning into oneself—
    remaining
        where one was but
    taking a conveyance
        for doing so—only
    the conveyance is inoperative
            the bus is still.

    Or being transformed
           by a certain process
       but the process is
             inoperative—
   nothing remains suspended
            in a solution of pure water only
and one remains   just as one was
          against the process . . .

10/26/87

At whatever moment we choose to inspect the owl
the owl remains
projecting itself
across a continuum of moments.

And the stubbly field
where the lonesome oak remains—
remains.

And every granite stone of it
throws itself
across a continuum of moments.

10/28/87

We peel away the cloak of night
and glance at the darkest stone.
It stands in the stubbly field.

We peel away the cloak of night and view
the moment of the oak.

Its bronze leaves shake in the wind.
A shudder of sound ascends.

We peel away the cloak of night
and peer upon the purchase of white owls.

They clutch the branch.

We peel away the cloak of night and glance
at the moment of ourselves.

The storm has broken a limb from the oak in the night.

11/3/87

A little crystal globe
rolls towards some white moon.

And in it a species creature
sits in lattice rungs.

And when the waves of moonlight sweep across it
the globe will lose its smallness and become
the universal sphere of abstract stone.

The creature will lose its species and become
all of us—

all of us will drift toward that white moon
in whose waves of light we lose our nature.

In the Sad World waves of moonlight pass across
the strings of precious stones our moments seem.

The ruby light adrift across a neighborhood
loses the purity of the color it has.

The world grows brown or dense or sooty specks
appear in points of air.

Poems from The Sad World

In the sad world waves of moonlight pass away
and the darkened air recovers its rebeous purity

and the dreams go on
and the changes prosper

11/6/87

The elk mind moves its antlers
                flourishing the trees....

The little animals huddle in the weather.

      (People are weather deities, storming the world.)

They leave their minds in trees
       green in the wind
   and milky fibrations stream across the skies

and the white owls hold to their branches

and the black stones stand in the field.

11/9/87

Abstract machines: the grinding of mice
that build the house they eat

according to the laws that lose themselves
as the ratchets clatter ...

11/11/87

Metropolis "Ah"

To it all trucks return. Or can
return. They can be recalled there.

Wherever they travel their cargo
      across whatever blank places

    to Metropolis "Ga" or "Ka"
or to some confuted village along the clefts—

From Metropolis "Ah" a term is set to the movement.

And the elk mind flourishes—

It holds the city "Ah" in its changing antlers.

And the creatures forget their natures
        as they travel
      back across the route
to the root metropolis

They forget their house of mice
        at Metropolis "Ah"
    traveling the goods to fuel the towns—

They forget the long confuted towns along the clefts
    and the rules they rode to get there
    and the owls they used

traveling back to metropolis "Ah"
        forgetful
                longing

Or forgetful, pulsing with the bliss of it.

The town of jeweled pavements
        washed by that white moon

and the jewels transfered to bins
        in another town, glinting—

And the walls of the town, what are they now?
    All walk across the line they draw
        or pass like mist through open grates—

And the news they travel on conveyances
        coming from another town
    across degradable spaces—

The news reports the space it passes through.

A voice comes on
        and the edges of the speech it uses bleed—

Poems from The Sad World

The speech bleeds into not speech.

Particle by particle
        replaced by alien tokens

    grd   th   t
       k   n s   t   sh
d   um   uu   i

Syntax preserved behind a Vigil of Negation—

The wardens remain in their booths
        atop the walls
      fielding the trucks.

The universal animals
    retract
        to the space behind
            the speech they use
to the rules before the rules were set—

They sit at archaic tables
        with savage goblets
      posting the first generalities

. . .

And the moon wave passes o'er
  and the language thickens—

The beasts become ourselves
    ensconced in natures.

11/13/87

The first generalities —alone
in the broken
school . . .

(the broken pavement sumac spumes in the air)

It gets smaller . . . Beasts are huge
and walk across the yards . . .

Charles Stein

People diverse from the analyses that cross them
abuse the thought they find themselves inside of

losing themselves among the thoughts they find

confuting the worlds.
                        Every person
confutes a different world
and wanders, blissful, desolate

a retract of the apparatus map
                that spots them
   a declension of the
         beast without a nature

   searching for a   Door
            to get back Out
    or up
     or in
            or forth or onward—

  to the space that marks
         the direction of direction

home to metropolis "Ah"
      general

            without foundation

Christer Hennix

$$\breve{E}[\eta]$$

(Text Of Order Type η)
[Detail]

from The Yellow Book
[Tractatus De Intellectu Emendatione Antiqua]

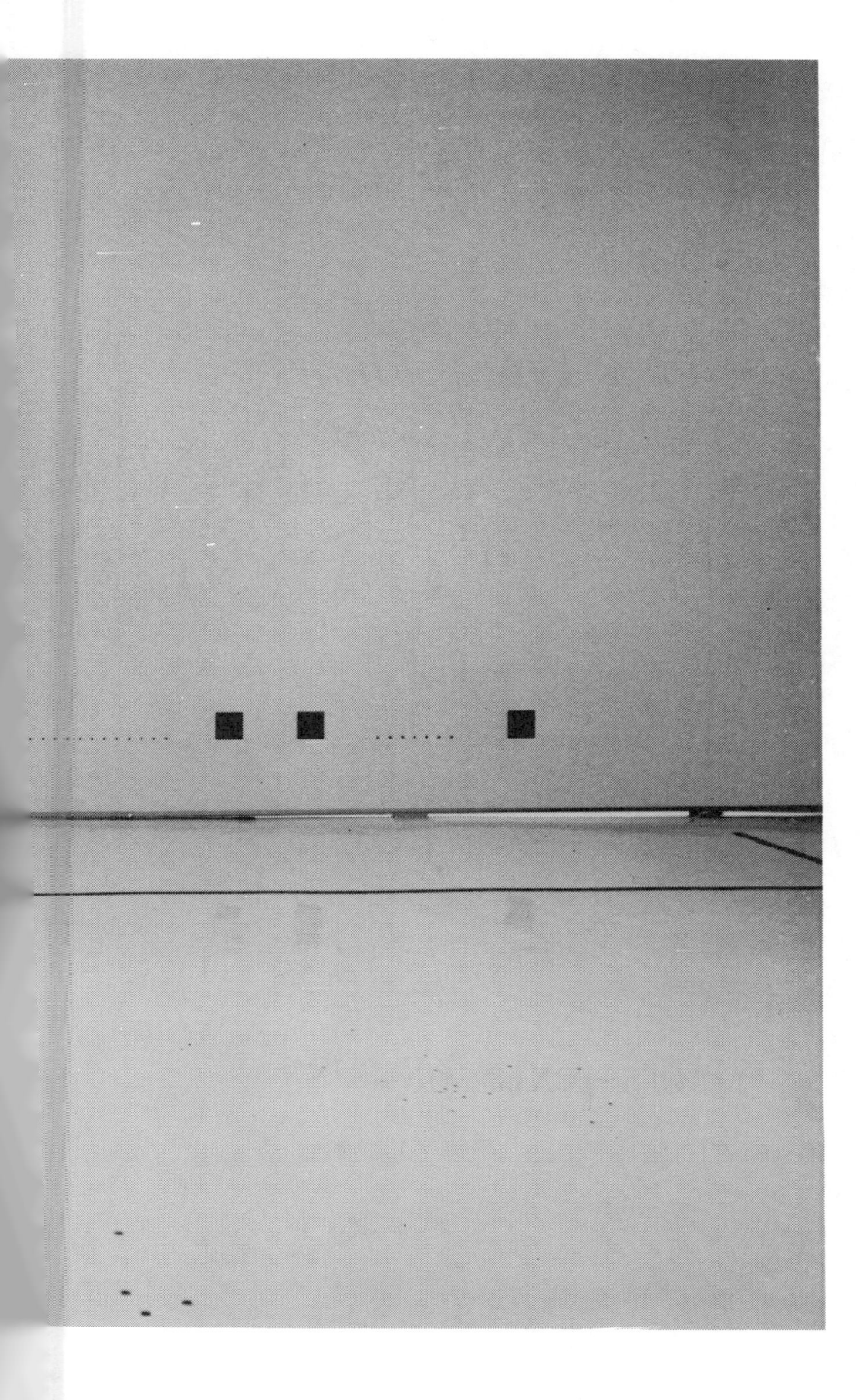

Finis Universatum:
Philosophy As Art /
Philosophy As Notation
II

— — — — — XX — — — — —

Hors-texte '68—'88

Abstract Tactical Configuration (Complete view 22 X 76 — Moderna Museet, Stockholm)

»— — — — —«[1]

$$\xi\upsilon\nu\grave{o}\nu\ \delta\grave{\epsilon}\ \mu o\acute{\iota}\ \dot{\epsilon}\sigma\tau\iota\nu,$$
$$\dot{o}\pi\pi\acute{o}\vartheta\epsilon\nu\ \ddot{a}\rho\xi\omega\mu\alpha\iota\cdot\ \tau\acute{o}\vartheta\iota\ \gamma\grave{a}\rho\ \pi\acute{a}\lambda\iota\nu\ \ddot{\iota}\xi o\mu\alpha\iota\ \alpha\tilde{\upsilon}\vartheta\iota\varsigma.\ \text{Bl. Fr. 5}$$

---

[1]Footnote 1 page 284

The Yellow Book

Preliminary Concepts

It is the tradition of Philosophers since Vedic times to inscribe Being as a term in a *text*.

By this tradition, Being has meaning *only* as a part of a term of a text. As a part of an inscription.....

I am considering terms which have paths through the following (already 2-large) succession of texts

..... Ṛg Veda ..... B1. (ΠΑΡΜΕΝΙΔΟΥ ΠΕΡΙ ΦΥΣΕΩΣ) ..... Hekigan Roku ..... Over de Grondslagen der Wiskunde ..... Tractatus Logico-Philosophicus .....

in order to explore if, at all, they will have another successor of a text of the unwritten doctrines, *agrapha dogmata*,* or, if the last texts have already been written but remain invisible in a historical era of oversymbolization and oversimplification.

---

*Project B1.29-30.

It is not that I am looking for something important which has been written down somewhere in the above sequence of classical texts—I am *not* trying to be instructed by these texts, but, rather, I am interested in the dual relation of attempting to instruct these last texts *themselves*—by subjecting them to the primary processes of *semeiosis*. In particular, I am contemplating the sources of/for *homosemeiosis*

$\Sigma \longrightarrow \Diamond$ —the Creative Subject instructs a text τ.

The dual relation is more familiar (from the theory of algorithms).

$\Diamond \longrightarrow \Sigma$ —the Creative Subject is instructed by a text τ',

although the following fact is apparently quite unfamiliar, *viz*, that, in general, *neither τ nor τ'* are ever present or real objects.

Remark. The concept of (an ever present) time or reality is intrinsically an *ethical concept*. The present or the real is sometimes *permitted* and sometimes *forbidden*, and, if neither, can be transgressed by a *new modality of freedom.**

Last In-First Out. The real task in sorting out a sequence of classical texts is that of *eliminating* the latter or at least *replacing* them by a text

---

*This is the site *near* the topos of an *initial model subject*, viz, the topos under which one attends an investigation by *one's own tactics of attention*, *only*.

It is under such "toposes" that the Creative Subject realizes a complete sense of (perceptual or cognitive) freedom.

of radical translation [as, for example, the Sophocles translations produced by Hölderlin.] The later the original text is, the easier seems the task of superseding it. For that reason, a lot of work has been spent on Brouwer's and Wittgenstein's writings. —But even when so simplified, the task remains very complex and only *partial results* have been obtained as I have attended this investigation with *my* (prefered) tactics of attention.

* * *

Before entering the text, something else has already entered. The allowance or permission to gaze at a linear collection of signs the meanings of which are sometimes even forbidden—or, else, impossible to understand—has entered as a distant premonition before an uninterrupted flow of signs focused by the pages of a text which is condemned to continuously lose its authorship or be abandoned by its reader—all signs of certain distress (of certain veridical ambiguities).

Freedom—as opposed to certain distress. The concept stands in need of further investigation.*

---

*A second initial model subject would contemplate the *ethics of meaning*. The subject may begin by regarding the following set of definitions, *viz*:

i. The *positive meaning* of a sign is determined by what the sign is *allowed* to signify.

ii. The *double-negation meaning* (or *weak* positive meaning) of a sign is determined by what the sign is *permitted* to signify.

iii. The *negative meaning* of a sign is determined by what the sign is *forbidden* to signify.

iv. The *possible meaning* of a sign is determined by what *it is not forbidden* for the sign to signify.

[These are the *active meanings* of an arbitrary sign [to be distinguished from its *passive* or *lexical meaning*(s).]]

Digression. *What Is (Not-) Not An Axiom* (Frege).

For Frege, a proposition P is a *function* P(X) which takes *itself* as argument, P(P), and takes the truth-value $\top$ if P is *true* and the truth-value $\bot$ if P is *false*.

If P expresses an *axiom*, then P corresponds to a "constant function" $\top$ (with due abuse of notation) while its negation, $\neg$ P, corresponds to an *anti*-axiom i.e. a constant function $\bot$. The *functions* P must not be multi-valued or undefined, else P is logically equivalent to a *potential contradiction* such as

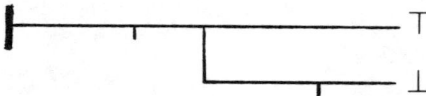

i.e. [ignoring intuitionistic doubts about double-negation elimination]

$$\top \mid \bot$$

in the notation of Tractatus Logico-Philosophicus.

However, it ought to be stressed that Frege somtimes considered an arbitrary proposition P as a *partial* recursive function—simply because P's truth-value is sometimes (intuitionistically) *undefined*.

That is (again resorting to Wittgenstein's compact notation),

$$P \mid . \; P \mid P$$

corresponds to a logically indefinite (or undecidable) proposition (contradicting the assumed completeness of the logical symbolism).

Digression cont. *Remarks On The Aesthetics of The Exterior* [A possible Kunstsprache; **Hors-texte**]

The exteriority of human existence reveals itself already by the operation of negation which signifies the absence or otherness of a modality of existence incommensurate with the present.

Not-*A*

signifies that *A* does not signify what is *now* (because "now" is *now* signified by "**Not-*A***").

Not-*A*, when signifying, signifies an aspect of what is present, of what is *now*.

Although the operation of negation brings forth an instantaneous temporal division between *now* and *not*-now, the otherness of this temporal *not*-now represents only a single layer of the exteriority with which human existence is universally confronted.

On the other hand, even when the logical sign for negation effectively brings agents of cognition to the exterior of their existence the aspects of the world revealed by an act of mental exclusion remain distinct from those aspects where the exteriority appears as a continuous action of successive intrusions by (an uncensored) reality.

Furthermore, the exteriority revealed by a *negationless* proposition

$$A'$$

also constitutes an infinite barrier for an agent of cognition to be compared only with the one presented by the **not-A'**. Because, if the proposition

$$\textbf{\textit{not-}}A'$$

is not false now, then the negationless proposition A' does not signify a cognitive aspect of the present now—except as an unsolvable or regressing vacancy. "A' holds" means that **not-**A' is not presently satisfiable, i.e. that *now* "**not-**A' does not hold".

The paradoxical conjunction "A' and **not-**A'" signifies a cognitive collapse of temporal order, i.e., it signifies (a-)symmetrically that *now* is *not*-now. That what is present is absent—with the absence asymmetrically not always present. [The asymmetrical aspect of a diagonal argument).

[**Reminder on scholastisism.** In the medieval logic *via moderna* two conditions had to be satisfied before an affirmative proposition could be said to be true. *Id est*, it must signify things to be as they are and it must not falsify itself. By contrast, a negative proposition is to be true just in case either it signifies that things are not as they are not, *or*, it has a self-falsifying contradictory. This distinction between affirmative and negative propositions carry two important consequences, *viz.*, pro primo, "This is false" and "This is not true" have distinct truth-values when they are both *self-referential*. Hence, while all propositions which signify themselves to be false indeed come out false on this analysis, quite paradoxically all propositions which signify themselves not to be true are nevertheless all true. The reason herefore is, of course, that while an affirmative proposition must satisfy a double truth-condition, a negative proposition needs only a single truth-condition to pass as a judgement. *Pro secunda*, tertium non datur must apply to any pair of contradictory propositions, *id est*, exactly one member of such a pair must be a true judgement while exactly one member must be a false judgement. Concisely put, "This is false" is false because it (without mediation) falsifies itself, while "This is not false" is true because its contradictory falsifies itself—again, without any mediation. Notice that as a result, Tarski's theory of truth or its contemporary variants (Kripke *et al*) are hardly applicable since the Tarski-like formula "'T' P' iff P" no longer can be appealed to (even with the additional premise that P exists and that 'P' signifies P, for P may be the case when 'P' is false—as when P is "This is false". That is to say, it is now possible for a false proposition to signify that things are as they are!)]

[1]*Lazy Dot Theory*

The natural numbers $0,1,2,\ldots$ are constructed as binary strings $0,1,10,\ldots$ which are denoted in a formal system by the numerals $0,0',0'',\ldots$. For computational purposes $0',0'',0''',\ldots$ is often notated as $\mathsf{I},\mathsf{II},\mathsf{III},\ldots$. These classes will be denoted by $\mathcal{N}, \mathcal{B}$, N, and H respectively. Their ranges are indefinite only in so far as the interpretation of the notation "$\ldots$" is indefinite, "$\ldots$" means "left to the imagination of the reader (including the author)," so "$\ldots$" has a definite meaning in terms of the variation of imagination allowed for by the class of readers. But since the class of readers is not a definite number the range of variation of the imagination is left indefinite.

Thus, "$\ldots$" is a sign of a class whether it occurs between terms as in (i) $0,1,2,\ldots n, n+1 \ldots$ or within a term as in (ii) $11\ldots1$. In (i), moreover, the class denoted by the occurrence of "$\ldots$" between 2 and n is, in general, not the same class as denoted by its second occurrence, and, in (ii), there is no *a priori* reason for identifying the meaning of "$\ldots$" with either of the meanings assigned to "$\ldots$" in (i).

Inspection shows that all of mathematics including set theory and, in particular, the formal systems, can be reduced to a complete formalization of the use of "$\ldots$" in modern mathematics. (This reduction, by the way, is the very (inductive) essence of modern Intuitionistic Type Theory.)

Ultra-Black Chaotic Topology—Painting of a Disaster
(Complete view 4 IX—17 X 76—Moderna Museet, Stockholm)

It is important to be able to grasp the composition or "graph" of the mixed use of lazy dots. If, say, (i) is used as the semantics for $\mathcal{n}$, i.e. $\mathcal{n}$ is closed, *at least,* under the action n + 1, then the length lh(n) of a numeral $0^{|...|}$ where "..." denotes n – 2 |'s is a *natural number* n i.e. a 0 – 1 string of length $\leq {}^2\log(n)$.

[However, it is clear that closure under n + 1 is not *used* in order to obtain closure under lh(n). The computational resources for closure under lh(n) are contained in "..." in 0,1,2,...n,n + 1, in (i), i.e. in the interval 2...n which is the "same" as 3...n + 1 relative any given interpretation of "..." (which is all we shall use)]

$\mathcal{n} \to$ H. *First* we speak about the natural numbers and *then* about the numerals which will denote them. The natural number 2, when denoted by the numeral 2, will also be denoted by the numeral 0''. Since in metamathematics everything is confined to a *finite* text T, it *suffices* to use the usual arabic numerals to indicate any subobject of T itself (as Gödel realized in his paper of [1931]).

N $\to \mathcal{n}$. Then numerals come in such plenty that all natural numbers which were imagined to take the place of the occurrences of "..." in the semantics chosen for $\mathcal{n}$ now are denoted in a consistent way by the numerals imagined to take the place of the occurrence of "..." in N.

According to Weil, Hilbert was of the opinion that a metamathematical capacity for imagination became illicit when it excluded certain commonly intuited limits. Thus, Hilbert ruled out as possible metamathematical terms those which exceeded 10^{12} discrete atomic steps by their "shortest" construction. What Hilbert wanted to emphasize here was the *perceptual* basis of metamathematical certainty as opposed to Brouwer's psychological or imaginary basis for the foundations of mathematics. Thus 10^{12} may be replaced by any larger but *fixed* constant relative which the possible numerical terms are defined. It should be within the spirit of Hilbert's program that the value of this *program constant* increases as new computational powers are developed. So understood, the solution to the 4-color problem is in perfect harmony with the modern spirit of *Hilbert's* program.

★ ★ ★

Hekigan Roku Calligraphy
Title page from an abandoned score (The 100 Model Subjects of Hekigan Roku, 1971—1976)

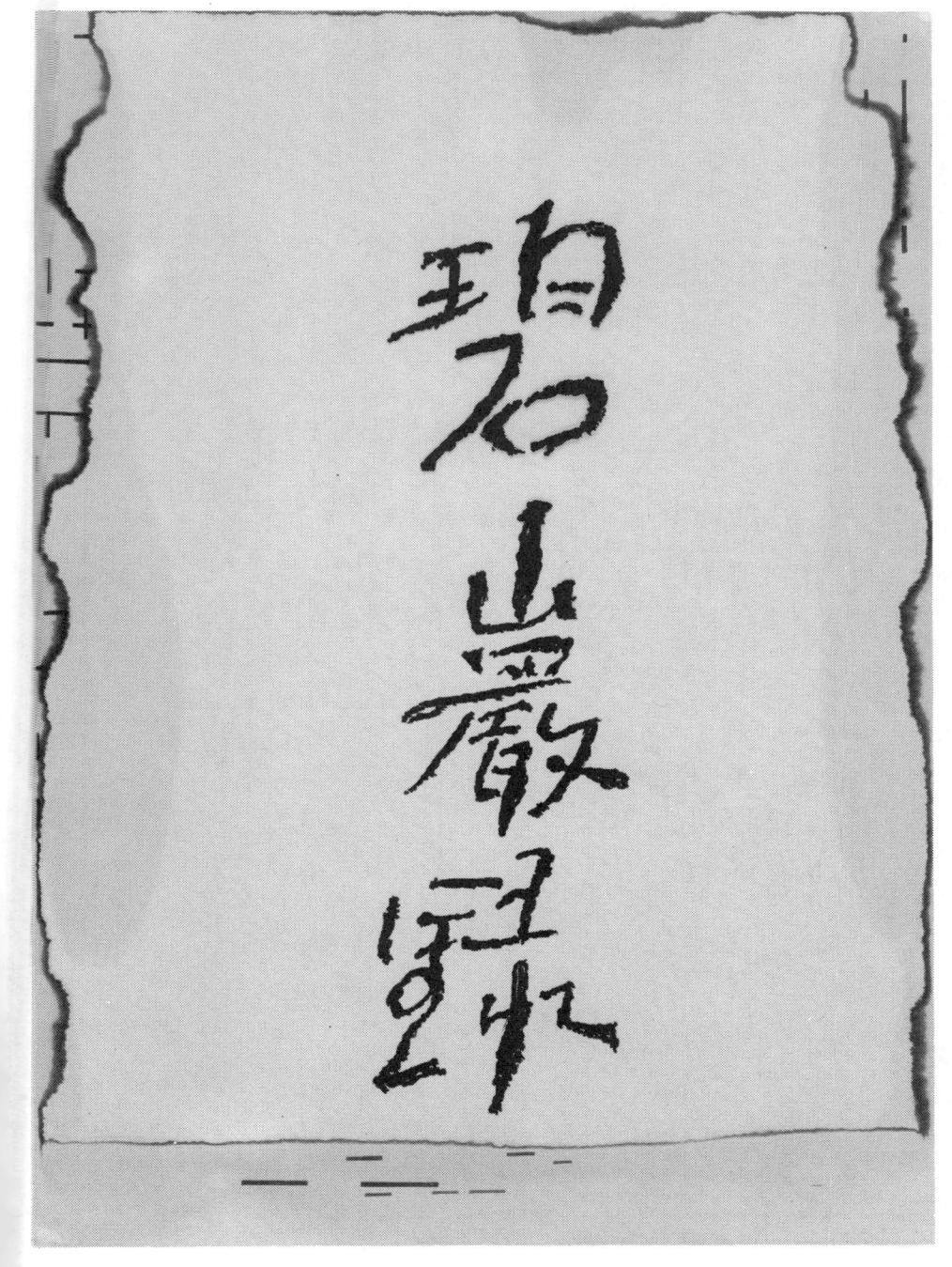

Two pages from the 100 Model Subjects of Hekigan Roku

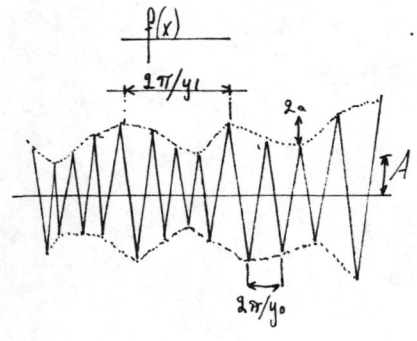

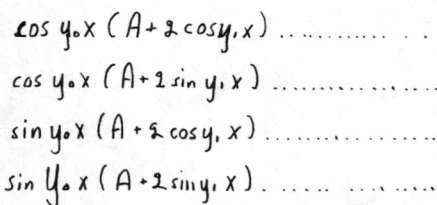

$\cos y_0 x \, (A + 2\cos y_1 x)$ ....

$\cos y_0 x \, (A + 2\sin y_1 x)$ ....

$\sin y_0 x \, (A + 2\cos y_1 x)$ ....

$\sin y_0 x \, (A + 2\sin y_1 x)$ ....

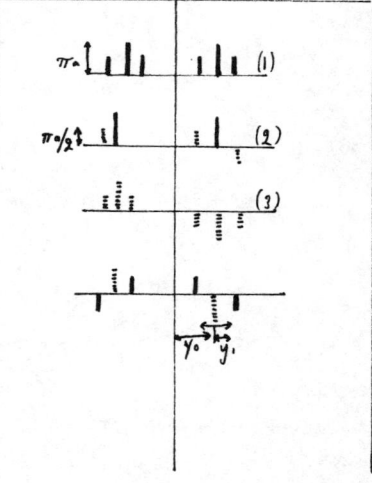

古譜尺八拍十聲指

MODEL SUBJECT NO 2

Jō-shū's 'The Real Way not Difficult'

Jō-shū is the Scholar of this subject. His personal name was Jū-shin. The name Jō-shū comes from the name of the temple at which he resided. He was a native of North China, but learnt Zen in the South, and did not return to the North till he was sixty-one years of age. He escaped the severest persecutions of his times because he was then living in retirement, and he lived on to the good old age of one hundred and three. He died in 897.

Jō-shū's words in the Main Subject, below, are based on the works of Seng-ts'an (Sō-san in Japanese), who was the third Chinese Patriarch of Zen. Seng-ts'an, died 606, is the author of the famous *Records Concerning Belief in the Mind* (*Shin Shin Mei*), and the opening words of this Main Subject below, 'The Way is not difficult, It dislikes the Relative,' are a direct quotation from the *Shin Shin Mei*.

INTRODUCTORY WORD

The universe is too contracted. The sun, the moon, the stars, the constellations (shining) simultaneously, are too dark. And even if the staff of correction comes down like drops of rain, and the scolding voice comes like claps of thunder, by no means do these provide any point of coincidence with the main subject of study—the Absolute Truth.

All that all the Buddhas of the three worlds could do was to understand it themselves. Even all the Patriarchs of the ages could not exhaust the explanation. The one whole enormous store of the Scriptures does not exhaust the explanation. Even clear-eyed robed monks have not fully understood the meaning of their own vision (accomplished their own salvation).

Having reached this stage, what should be done to obtain more profit (i.e. fuller knowledge)? To mention that name 'Buddha' is to drag yourself in the mud, to girdle yourself with water. To mention the word 'Zen' is for the whole face to show shame. Advanced students (Bodhisattvas) do not wait to be told this. Students following with their beginners' minds should immediately set forth to further study.

Interpretation of the Above

En-go introduced the subject by saying: The universe is too contracted and closed in for it to express the Absolute. The sun, the moon, the stars, the constellations, even when shining all together are no better than complete darkness for the purpose of revealing the Absolute.

Yes, and even if the teacher brings down his staff of correction, blow on blow, like rain in a storm, and even if he scolds with loud voice, like great claps of thunder, indeed whatever methods of instruction he may use they can have no application to, nor will they accord with, any real means of advancing along the way towards the Absolute.

Furthermore, even if all the Buddhas of the three worlds (past, present and future) were to try to explain this matter, all they would be able to do would be to understand it in their own selves. The whole line of Patriarchs, too, cannot rightly expound it. The great store of the Scriptures (twelve thousand volumes!) does not provide the explanation, and experienced monks who have had the vision have found the comprehension of Truth too hard for them to express.

Such being the case, what can we ourselves do to apply ourselves to this study? Even to talk about the name Buddha will make us feel as though we were being held back, dragging our feet in mud, or constrained as if we were being hindered by struggling along thigh deep in water. Advanced students, of course, who have practised meditation for a long time, do not have to wait to be told this. The minds of those who are still in the elementary stages must go on with their study with great sincerity.

MAIN SUBJECT

Attention! Jō-shū spoke to the assembly and said: The Real Way is not difficult, but it dislikes the Relative. If there is but little speech it is about the Relative or it is about the Absolute. This old monk is not within the Absolute. Do you value this or not? At this point there was a monk who said: If you are not within the Absolute how can you assess its value? Jō-shū said: Neither do I know that. The monk said: Your Reverence, if you do not yet know, how is it that you say you are not within the Absolute? Jō-shū said: You are effective in your questioning. Finish your worship and retire.

Interpretation of the Above

Jō-shū spoke to the assembly and said: The Real Way is nothing other than the Absolute. It was Seng-ts'an who told us that there is no difficulty about the Real Way, except that it is not expressible through the Relative. It dislikes the Relative. No matter how few words you may use in trying to explain this matter, you will soon find yourself talking about the Relative and the Absolute, and you will find yourself talking about differentiation. I myself am not captivated by the idea of the Absolute, so as to divide the real way into Absolute and Relative and to value the former very highly. And what about you? Do you distinguish between the Absolute and the Relative and value the former very highly?

Here one of the company spoke out and said: You acknowledge that you do not value the Absolute as highly as you would wish, but if so does not that mean that you are in some measure detached from the Relative? If you neither understand the Absolute nor remain detached from the Relative how can you assess the value of either? To this Jō-shū replied: I do not even know that. The monk said: If Your Reverence has no knowledge about these matters how can you even say that you do not prize the Absolute above the Relative? Jō-shū said: You are skilful in asking logical questions. Your logic is good. It is more important for you to go and do your worship than to argue logically. You will learn more in that way, so go off to your worship and retire.

APPRECIATORY WORD

'The Real Way is not difficult.' Words accord. Speech accords. Though it is One it has many aspects. Even if it be two, it is not only of double aspect. On the far side of the sky the sun rises, the moon sets. Outside the window the mountains are distant, the water is cold. Even though the skull's consciousness has ended why should joy have ceased? The decayed tree's dragon-moan may have been stopped, but it is not yet dried up (dead). Difficult, difficult! The Relative and the Absolute. Do you yourselves ponder them.

Interpretation of the Above

Seng-ts'an's words, 'The Real Way is not difficult,' are correct. Our very words, our speech accord with this. Through them the Real Way may be revealed. But what did Jō-shū mean when he said: If there is a little speech, if there are a few words they are about the Relative, they are about the Absolute? When unenlightened men talk about the Real Way, is what they say true? Rather would it appear from what they say that there is some one thing which might be called the Real Way. But in truth the Real Way is an inexpressible thing. It is of course One, but it has many aspects, nor is it to be seen as only two-sided, the Relative and the Absolute. It can be looked at from ten thousand directions.

Look out of your window at the world around you. There will be the sun rising on one side of the heavens, and the moon setting on the other. The sun and the moon, looked at from the point of the Real Way, are not necessarily two, neither are they necessarily one. They only appear so when looked at from the point of the Relative. Again, look at the distant mountains and the cold water outside your window. What relationship have these to one another? Are the mountains distant because the water is cold, or is the water cold because the mountains are distant? Truly distance and coldness are two and yet not two. They are one and yet not one. The Absolute and the Relative are exactly harmonized.

If a skull is found lying about, men say that consciousness has been cut off, but surely emotions like joy, anger, are still active. A decayed tree may seem to be dead, and yet when the wind blows we may hear it moaning like a dragon. The tree seems to be dead, but life itself is not extinct. It exists somewhere. From the point of the Real Way life is not necessarily life, nor is death necessarily death. Difficult, difficult! What is truly difficult is to express these truths in such a manner as to make them intelligible. That is why you must ponder them carefully.

Toposes
and Adjoints
(excerpts)

Main Subject: *Boundary Operators.*

A *boundary* is a *constant event* of a space. If the interior of the boundary is also a constant event, i.e. an event with no internal collations except for those which are constant, the space is also called a *constant space*.

The exterior of a boundary event is called the *outer event horizon* of its space, while its interior marks the *inner event horizon* behind the event itself.

Def.
A boundary. A distance. Def.
A boundary. A limit. Def.
A boundary. A reflection. Def.
(A motion without velocity.
My own diagonalization.) Def.

Space is a quality of mind, an intension, i.e. a modality of consciousness.

Space as a mode of mind, *not* as a definition — the confrontation between corrupt and enlightened cosmic matter — as a formation of acts of attention, as an array of signs, as a homotopy between an image and its mirror images.

Space is inseparable from the array of signs with which it is designated.

Undifferentiated space is described by its empty array of signs.

[Composition of Mind — Composition of a Space, indexed by its Boundaries.]

Signs are boundaries. They *presuppose* space!

Each formula expresses a possible boundary.

Functions and operators are generators of boundaries. Course-of-values — traces of boundaries.

The Yellow Book

Main Subject: *Pre-Socratic Set Theory*

a ∈ A ⇔ **a** is a program for the thought (intension) **A**

A *"school"* is a structure which contains *"classes"* some of which are called *"sets"*.

Schools create thoughts via their classes. They distinguish themselves by their syntax. In this sense, schools breed the trees of knowledge by the proliferation of classes arranged into hierarchies of (epistemological) domination.

Definition. Pre-Socratic classes exist previous to the formation of *all* their elements.

[***Example***. The point-species of a locally infinite space (of infinitesimal distances.)

Intuitionistically, a class symbolizes a *"species"* or *"type"* of propositions.

a ∈ A

a is a proposition of *type* **A**, or, **A** is the species of proofs **a** of **A**.

A set is a class together with a proof that it is determined by all its elements.

Therefore, the type T of *all* types is *not* a set.

A "small" type T corresponds to a "large", i.e. infinite, proposition.

Its members are "small". i.e., finite propositions, each proof **a** branching only finitely many times.

A formula is a possible proposition. In general, proofs are complex propositions since the complexity of the proposition proved cannot exceed the complexity of its proof. Atomic propositions are their own proofs. I.e., if A is atomic, then the type [A] has exactly one member when A is true, and is empty when A is false.

Thus, types arise in processes of analytic continuations of symbolic spaces, the general continuations consisting in assigning each proposition A a *type* [A] so that each false proposition is assigned the *empty type* [ ], and if A is a true proposition, then [A] is the *species of proofs* of A. That is to say, this assignment of types determines the meaning of a proposition in terms of its possible proofs.

Language forms a space divided by its propositions.

Theory of Mind = Theory of Boundaries of Being **Def.**
Logic = Theory of Boundaries of Cosmic errors **Def.**
Topological Geometry = Theory of Boundaries of Space **Def.**

The earliest results in the philosophy of thought showed the three-fold impossibility of

      i. enumerating space—
     ii. alphabetizing space—
    iii. geometrizing space—

from which the impossibility of (completely) axiomatizing space clearly follows.

. . . . .

Being forms a site of boundaries.
Its existence describes the
Boundaries of Being.

Incessantly,
Being describes boundaries,
Its incessantly orbiting limits.

The substratum of a boundary
is called *space*.

Geometry is the
syntax of space,
A cosmic phase-shift
A site of condensed cognitive
Matter

Clearly, a logical type [A] is a mixed bag of logical evidence, the extension of the mix being a function of the scope of the notion of proof under contemplation.

More precisely, a (non-empty) logical type [A] is a logical manifold which has the symbol A at its root, each main branch of the manifold representing the structure of a possible proof of A.

An immediate metalogical observation is that, because of over-alphabetization [overload] there must be true propositions which cannot be assigned any type at all. Propositions without types are *non-stratifiable* propositions.

If $\mathbf{b}(\epsilon \mathbf{A})$ is $\neg \mathbf{a}$, i.e. a proof of $\neg A$ (not-A), then $\mathbf{A}$ is said to form a *"diagonal class"* with $\mathbf{a}$ the symbol for its *"diagonal formula."*

A great deal has been written since Frege on diagonal definitions occurring in a *system of definitions*. It is known from (academic) folklore, that the following provocative picture of diagonalization was formulated by Wittgenstein sometime in the mid-30s (after Kreisel):

> *Suppose we have a sequence of rules for writing down rows of 0 and 1, suppose the pth rule, the diagonal definition, says: write 0 at the nth place (of the pth row) if and only if the nth rule tells you to write 1 (at the nth place of the nth row); and write 1 if and only if the nth rule tells you to write 0. Then for the pth place, the pth rule says: write nothing!*
>
> *Similarly, suppose the qth rule says: write at the nth place what the nth rule tells you to write at the nth place of the nth row. Then, for the qth place, the qth rule says: write what you write!*

Inevitably, the diagonal argument (however cheaply) brings out the *incompleteness* of any formalized notion of proof theoretic studies, *unless*, that is, they are equipped to deal with formal concepts of *negative evidence* which are available *independently* of *any* concept of proof (formal or not). The following observations are due to the *ultra-intuitionistic studies* in the foundations of mathematics.

Summary. Typical Example Of A Working Definition.

$\mathbf{A}$ is a type (set) ($\mathbf{A}$ is a formula).

A is a type — $\mathbf{A}$ **type** ; $x_1 \in \mathbf{A}_1, \ldots, x_m \in \mathbf{A}_m$

$A = B$ equal types — $\mathbf{A} = \mathbf{B}$; $x_1 \in \mathbf{A}_1, \ldots, x_m \in \mathbf{A}_1$

$a \in A$ object of type A — $\mathbf{a} \in \mathbf{A}$; $x_1 \in \mathbf{A}_1, \ldots, x_m \in \mathbf{A}_m$

$(a = b) \in A$ equal objects of type A — $\mathbf{a} = \mathbf{b} \in \mathbf{A}$; $x_1 \in \mathbf{A}_1, \ldots, x_m \in \mathbf{A}_m$

.....

The sign of Being in Philosophy is in the signs of a text while the sign of the text is a sign of non-Being of Being in Philosophy and its texts. [Iterated diagonal formulae]

Being is a term of a text. Its meaning is *only* that of a term of a text. Yet, paradoxically, Being may be enriched as a term of a text.

What exists, what surrounds Being, is a text.

It is through the action of the text that Being as Thought becomes a Form in many variables.

It is the action of the many variables of Thought which is the source of the many forms of Being.

Thought is a Form in many variables. E.g. ...x...y...z... is a formal expression of a form of a Thought in which enters at least three variables denoted by x, y, and z.

A formal expression of a Form is a sign of limits as Being itself is a sign of limits.

A text is an arrow along which a natural transformation of Being into co-Being occurs.

A text presupposes space.

Space exists as Being.

Being generates space.

Space carries geometry as Being carries space.

The geometry of a Form in many variables is called a *site*. It is formally denoted by a category together with a topology.

The *analytic continuation* of a form in the algebraic geometry of a site is called a *pre-topos* while its *analytic completion* is a *topos*, a continuously variable natural transformation of the intrinsic geometric logic of the algebraic geometry contained in every possible analytic continuation of a Form of a site.

The geometric logic *intrinsic* to the geometry of a Form is not classical but *intuitionistic*. This can be seen in the geometric constructions corresponding to the logical constants and in particular, the quantifiers, including negation.

The notion of a topos effects a natural transformation between a geometric space and a logical space intrinsic to the geometry of the space.

The geometry of space is a perception of inferences in logical space, [The geometry of Thought is a geometry of Being.]

Being defines a space together with an action.

The Alternative Interpretations Of Type Theory.

1. A is a proposition (A is true) (**a** is a proof of the proposition A).
2. A is a problem (A is a solvable problem) (**a** is a program for the problem A).
3. **a** ϵ **A** — **a** is a (mental) program for the thought or intension **A** (**A** is a workable idea).

Remarks on the alethic realities. As soon as an event is perceived by Σ, its arrival is *"organically" real* and the sentence which expresses this arrival is *"epistemically" real*. For example, any arrival of a perceived collation is organically real and any such collation may be included in the text of the epistemically real sentence which expresses the perception.

A situation **S** is *"organically real"* (*"epistemically real"*) as soon as each sentence to be accepted by the sense of **S** is *"organically"* (*"epistemically"*) real and any of its textual collations is perceived. In particular, if **S** is a collection of sentences and textual collations, its organic (epistemic) "reality" shall be that of each of the sentences together with the perceptions of the textual collations of **S**.

If a sentence on the presence or absence of an object (in **S**) is organically or epistemically real, the presence or absence of the object (in **S**) shall be called, respectively, organically or epistemically "*real.*" In general, only logically indecomposable objects or events are considered as the content of "perceptions" or what is "perceived", while more complicated objects are to be deduced from the reality of decomposable objects by logical means, which are studied by a specific branch of (ultra-intuitionistic) logic, viz, the *Reasoning Theory*. When the Reasoning Theory is available, "exhaustions" are considered as indecomposable objects when the means to obtain them by the *Conjunction Theory* fails.

I now indicate four ways, $\alpha-\delta$, by means of which a sentence on the absence or presence of an object **e** or a class **E** of objects may be recognized as real in **S** *independently* of the Reasoning Theory.

δ. If any object present in **S** is discerned from **e** and all objects present in **S** are exhausted, then **e** is not-present or "*absent*" in **S**.

β. If any object present in **S** is discerned from any object of the kind **E**, and all objects present in **S** are exhausted, then no object of the kind **E** is present in **S**, or, equivalently, all objects of kind **E** are "*absent*" in **S**.

Being is a sign of a site of limits.

A text delineates the geometric limits of Being.

To ask what is a text is analogous to asking what is space.* I.e., again, there is an analogy with the continuum. The space of the text is yet another replica of the continuum problem, i.e., problems of thought connected with the existence of infinitary or ω-forms of thought ...x...y...z... in many variables x,y,z,....

---

*No part of a text avoids defining a space whose intrinsic geometric logic is not intuitionistic. Thus, as non-euclidean geometry was not immediately asserted by the unguided perception of pure space so non-classical logics were not observed to occur in the passages from finite to the infinite in Thought.

Now, it is well known that at least since the time of Plato, a course in *advanced philosophy* has always taken the knowledge of geometry for granted. Riemann, Frege and recently Lawvere have in different ways approached fundamental epistemological questions by means of purely geometrical considerations. This situation has led to a Model Subject with the following main subject: The geometry of Thought in the becoming of a geometry of Being.

γ. If **e**, or any other object **b** in **E**, is *alien* to any of the ways of recognizing an object as "present" in **S**, then **e**, or any object **b** in **E** is not-present ("*absent*") in **S**.

δ. If the arrival of Oc_e of an event **e** is absent in **S**, then **e** is *non-arrived* in **S**.

On the basis of α–δ one may now specify the following

Principle of Negative Extension

Let **S** by any real situation and all objects of the kind **E** absent in **S**. Then the situation **S**, obtained from **S** by adding the sentence that "*all objects of kind* **E** *are absent in* **S**", is real.

The principle of negative extension (**p.n.ex.**) has been considered by various authors throughout the history of logic. It is interesting to note that **p.n.ex.** can be used to obtain an axiom of "infinity" as was actually done by some scholastic logicians and, notably, by the Indian logician Dinnaga (9th century C.E.).(*Id est*,there exist ontologies with "infinitely" many *absent objects*).

. . . . .

We have now developed some means in order to declare some situations "*alethically real*" since we are able to take into consideration some *sources of sentences*

$$\Diamond^\chi_\S A_1 \quad M_S A$$

from which an ascent to an alethic reality can be made by the reader. Viz. recall that some sentences $\Diamond^\chi_\S A$ are already available since every prepared atomic action **A** is supposed fulfillable i.e. is *organically possible* in any organically possible situation. Further, since we demand that all prepared atomic actions must be allowed, the corresponding deontic sentences **P** A must be accepted as must the epistemically corresponding sentences **M** A. Hence, the following two definitions belong to the Modality Theory.

I. A situation **S** is alethically *real* if each element of it is accepted on the basis of a perception *or* a definition.

II. Each alethically real situation **S** will also be called a *possible* situation.

Note. $\Diamond$, **M** and **P** are formal symbols for *alethic, epistemic* and *deontic possibility*, respectively, where χ denotes a parameter varying over some set of "laws".

Christer Hennix

Model Subject W

$$\left\{ \text{Truth Set Of Measure Zero} \right\}$$

Introductory Word.
(4.093. 4.092)*

4.094. Ein Bild zur Erklärung des Wahrheitsbegriffes: Schwartzer Fleck aus weissem Papier. Die (Papier; die) Form des Fleckes kann man beschreiben (,) indem man für jeden Punkt der Fläche angibt, ob er weiss oder schwartz ist. Der Tatsache (,) dass ein punkt schwartz ist (,) entspricht eine positive—der, dass ein Punkt weiss (nicht schwartz) ist (,) eine negative Tatsache. Bezeichne ich einen Punkt der Fläche (einen Fregeschen Wahrheitswert), so entspricht dies der Annahme (,) die zur Beurteilung aufgestellt wird. (,) etc. etc.

Um aber sagen zu können (,) ein Punkt sei schwartz oder weiss, muss ich vorerst wissen (,) wann man einen Punkt schwartz und wann man ihn weiss nennt: um sagen zu können (:) „p" ist wahr (oder falsh) (,) muss ich bestimmt haben (,) unter welchen Umständen ich p („p") wahr nenne, und damit bestimme ich den Sinn der Satzes.

Der Punkt (,) an dem das Gleichnis hinkt (,) ist nun der: Wir können auf eines Punkt des Papiers zeigen (,) auch ohne zu wissen (,) was weiss und schwartz ist; einem Satz ohne Sinn aber entspricht gar nichts, den er bezeichnet kein Ding (Wahrheitswert) (,) dessen Eigenschaften etwa „falsh" oder „wahr" hiessen: das Verbum eines Satzes ist nicht „ist wahr" oder „ist falsh"—wie Frege glaubte—, sondern das (,) was „whar ist" (,) muss das Verbum schon erhalten.

---

*Cf. Tractatus 4.063

Interpretation of the Above
Finite Set of Measure 0 (Detail) 1969 (2-D Truth-Space) Moderna Museet 1976.

Model Subject No. –1 [Without Title]

Introductory Word [1-2.0121...]

1. The world is a form of thought.
1.1 The world consists of forms of thought, not of facts.
1.11 The world is determined by forms of thought and by them all being nothing but forms of thought.
1.12 Because the totality of forms of thought determines what is as well as what is not.
1.13 An exact world consists of forms of thought in logical space.
1.2 Every world breaks down into its constituent forms of thought. The world eventually breaks up in its atomic forms of thought—when the logical analysis has been completed.
1.21 The exact forms of thought form an independent set of forms, where an exact form corresponds to a form of an exact thought.
2. What is the case—the form of thought—is the existence of states of minds.
2.01 A state of mind is a combination of thoughts.
2.011 It is essential for thoughts that they should be possible constituents of states of mind.
2.012 In logic nothing is accidental: if a thought can occur in a state of mind, the possibility of the latter must be co-possible with the thought itself.
2.0121 It does not stand to reason that a form of thought shall adjust itself to a thought that exists independently of the former. If the thought can occur in a form, then that is something co-formed by the formation of the form itself. (What is logically possible is not just simply possible. Logic is occupied with every possibility, and all possibilities form its field of thought).

Just as we are completely unable to imagine spatial objects outside space or temporal objects outside time, so too are we wholly unable to conceive of a single thought without the possibility of its connections with other thoughts, without its occurrences within the boundaries of a form of thought.

If I can imagine thoughts combined and connected by forms of thought, then I cannot imagine the possibility which excludes every thought from at least one such combination.

.
.
.

LOGISCH-PHILOSOPHISCHE ABHANDLUNG

1* Die Welt ist alles, was der Fall ist.

1.1 Die Welt ist die Gesamtheit der Tatsachen, nicht der Dinge.

1.11 Die Welt ist durch die Tatsachen bestimmt und dadurch, daß es alle Tatsachen sind.

1.12 Denn, die Gesamtheit der Tatsachen bestimmt, was der Fall ist und auch, was alles nicht der Fall ist.

1.13 Die Tatsachen im logischen Raum sind die Welt.

1.2 Die Welt zerfällt in Tatsachen.

1.21 Eines kann der Fall sein oder nicht der Fall sein und alles übrige gleich bleiben.

2 Was der Fall ist, die Tatsache, ist das Bestehen von Sachverhalten.

2.01 Der Sachverhalt ist eine Verbindung von Gegenständen (Sachen, Dingen).

2.011 Es ist dem Ding wesentlich, der Bestandteil eines Sachverhaltes sein zu können.

2.012 In der Logik ist nichts zufällig: Wenn das Ding im Sachverhalt vorkommen kann, so muß die Möglichkeit des Sachverhaltes im Ding bereits präjudiziert sein.

2.0121 Es erschiene gleichsam als Zufall, wenn dem Ding, das allein für sich bestehen könnte, nachträglich eine Sachlage passen würde.

* Die Decimalzahlen als Nummern der einzelnen Sätze deuten das logische Gewicht der Sätze an, den Nachdruck, der auf ihnen in meiner Darstellung liegt. Die Sätze n.1, n.2, n.3, etc. sind Bemerkungen zum Satze No. n; die Sätze n.m1, n.m2, etc. Bemerkungen zum Satze No. n.m; und so weiter.

Interpretation of the Above (vertical *s*equence of dots)
The general form of the *natural numbers* is given by

$$N_W \qquad\qquad [0,\ \xi,\ \xi+1]$$

where 0 is a parameter for values of the *empty logical operation*, Λ.

Numbers are exponents for an operation.
All initial operations are logical operations.
The operation of picturing a fact (as a composition of arrows) is a logical operation.
By picturing the fact that an operation has taken place, by constructing its model, the *meaning* of that picture is now attainable as a new logical object, as a new collection of arrows.
There are three *parameters* in the number theoretic schema N_W.
0 is a parameter for values of Λ.
ξ is a (meta-)variable of quantification—the most general (logical) designation of a natural number.
$\xi + 1$ is the generic parameter for a fixed arithmetical *succesor function* with argument-values *restricted* to the range $N_\xi \subset N_W$ of quantifiers with ξ as the *only eigenvariabel* [cf. Frege's general concept of a (partial recursive) function].
All mathematical operations are logical operations.
Numbers don't occur in logic, they are a side-product—the exponents of operations.
Logical propositions say nothing, they cannot define the exponents of an operation.
Not even the exponent of the empty operation can have a logical name, since, if I were to choose one, $\emptyset$, say, then by "$\emptyset$" a dynasty of new arrows would follow in the wake of tracing the developing identifying chains, in tracing the ensuing compositions of arrows.
Mathematics does not take place within logic.—It occurs outside of logics, as a theory of models for formulae with quantifiers.
The general form of numbers is mirrored in the general form of *truth-functions*

$$L_W \qquad\qquad [\bar{p},\ \bar{\xi},\ N(\xi)]$$

which is the general form of the *proposition* P expressed as a sentence p [provided that the elementary truth-function N(of P) is at most ξ-ary].

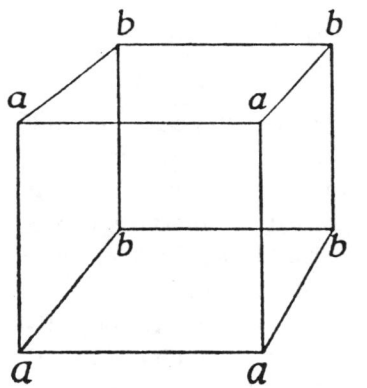 ≠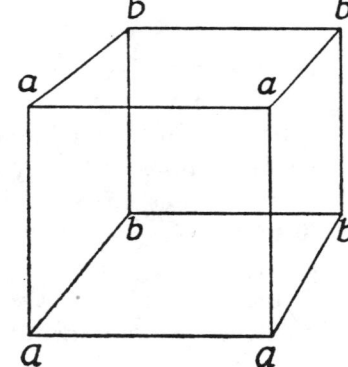

That is, out of the ξ elementary propositions p,q,r,..., an infinite sequence of truth-functions of elementary propositions is obtained by iterating the continuously variable elementary truth-function $N[\eta]$ infinitely many times with any previously obtained values as its arguments. I denote the so obtained sets of propositions by $\bar{p}_\xi$ and regard it as a *covering* of the space of propositions determined by L_W. [Indeed, $\bar{p}_\xi$ suitably defined may be considered as inducing a *Grothendic topology* on the language L_W. Although one will notice complications which will arise if the argument ξ accepts non-Archimedian integers, i.e. integers $\underline{\xi}$ such that

$$\underline{\xi} \neq 1 + 1 + 1 + \ldots + 1$$

i.e. such that $\underline{\xi}$ is *not* a natural number.]

For any choice $\bar{p}$ of elementary propositions p,q,r,..., suitable definitions of the set $\bar{p}_\xi$ represents an enumeration of *all two-valued truth-functions* uniformly built up around the elementary truth function $.|..$ Thus, the set $\bar{\bar{p}}_\xi$ of all sets $\bar{p}_\xi$ regarded as binary sequences (0-1-sequences) is in 1-1 correspondence with all infinite sequences of binary natural numbers regarded as exponents of (iterated) metamathematical operations over the propositions in the language of arithmetic. Once more, the totality of all paths taken through a set of all meaningful propositions is made to depend on the meaning of the *continuum problem*. [Needless to say, the following hypothesis, *viz.*

(CH) $\qquad\qquad 2^{\aleph_0} = \aleph_1$

(the *continuum hypothesis*), is *not* included in any totality of meaningful propositions but is simply regarded as one among Cantor's many wreckless statements — a verdict, by the way, which didn't win common acceptance until "after Cohen" when it was finally realized that the infinitary exponential term $2^{\aleph_0}$ *can be anything it ought to be*!

.
.
.
.

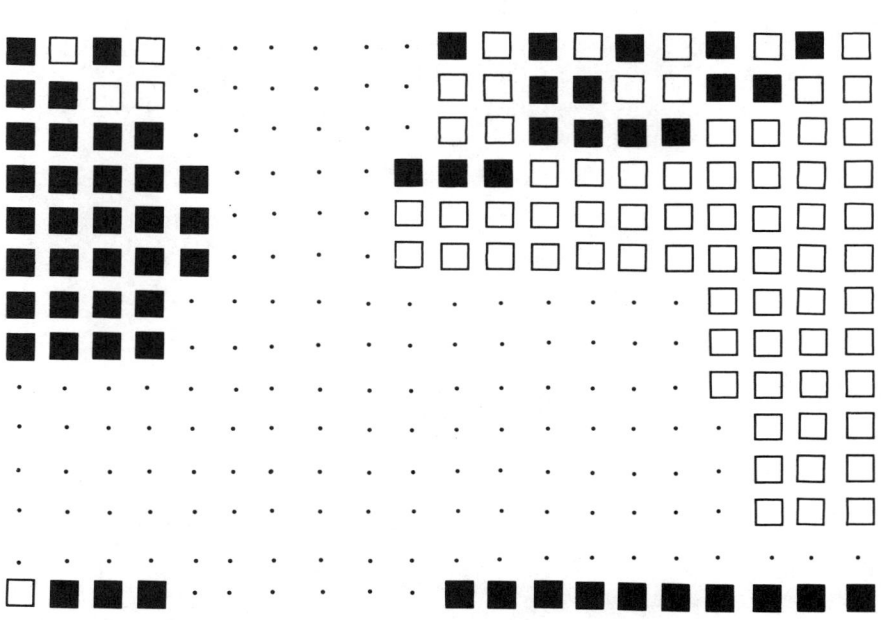

Christer Hennix

Main Subject
On The Metamathematics of Tractatus Logico-Philosophicus

Attention!

I shall only state this much as known:

> The clarity of a form of thought is a cosmic event, waiting to take its place among the elements.....
>
> Now, regard the well-known philosophical diagrams below—in the relative and in the absolute. Do you yourself ponder them!

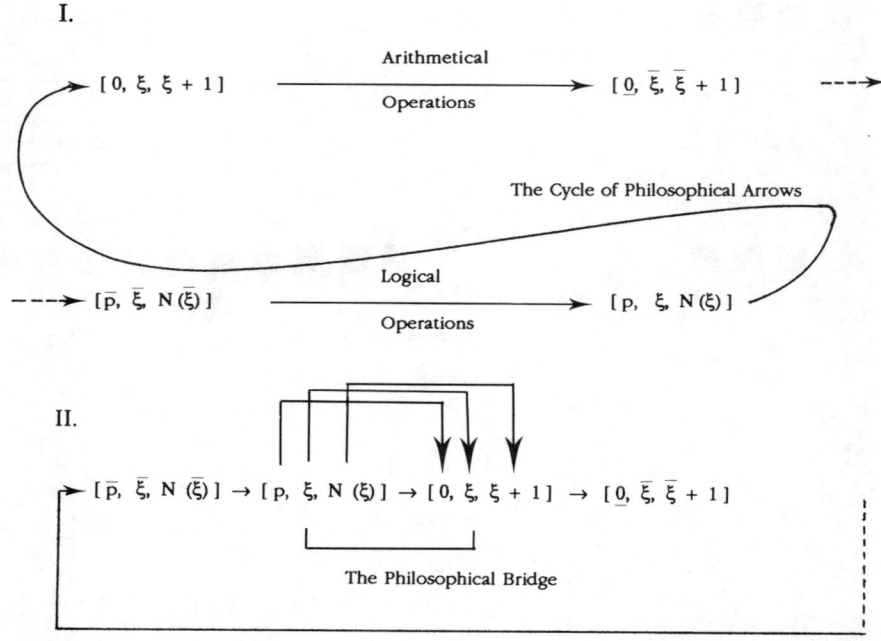

The Yellow Book

Interpretation of the above

Any particular development of arithmetic cannot be chosen on the basis of the propositions of logic alone.

A form of arithmetic, such as N_{WA}, is a form of *restricted quantification*.

That is, in order to indicate any particular arithmetic from the general form of natural number,

$$[0, \xi, \xi+1]$$

I need to specify the ranges of values of the parameter ξ.

The general form of arithmetic is

N_{WA} $\qquad\qquad [0^{\bar{\xi}}, \bar{\xi}, \bar{\xi}+1]$

Like the propositional variables, the arithmetical variables arrange themselves in *types*.

Numbers correspond to exponents of operations on propositions.

[Alternatively, numbers measure the quantity of mind's folly (cf. L.E.J. Brouwer).]

Arithmetical types are also called *"functional universes"* or *"natural number objects"*. For brevity, I shall designate them as "f-*types*" when determined by the stages (course-of-values) of an f-sequence $fx = \bar{\xi}$.

Remark. Regarded as a *category* neither N_W (nor N_{WA}) is a *discrete category*, i.e. not every arrow is obtained as id_N on some arithmetical object N.

I now recall that each arithmetical object $n \geqslant 0$ forms a category *viz as follows:*

    0 —*empty category*
    1 —*singleton category*
    2 —*binary category*
    .
    .
    .

At the limit of this series of abstract constructions stands the category **N** formed by a single object N in conjunction with infinitely many arrows from N to N (corresponding to the natural numbers, 0,1,2,3...).

Each arrow has the same domain and codomain, *viz.*, the — unique object N. Composition of arrows m,n, corresponds to *addition*, i.e. m o n = m + n:

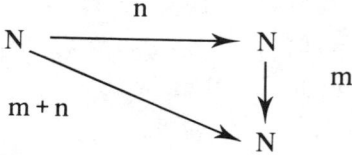

Furthermore, if the "philosophical bridge" between occurrences of the natural number objects N is indicated as $1_N = 0$, then

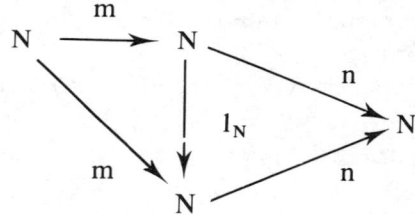

commutes, since 0 + m = m and n + 0 = n.

.
.
.

Appreciatory Word

(For the purpose of understanding the text below recall that both the concept of formula and the concept of arrows generalize to the concept of *diagram* (Topos Theory). Furthermore, the latter concept generalizes to the concept of objects such as (calligraphic) characters and other complex signs. On the other hand, it has been known since the late 60s (folklore) that *Howard* had secured a proof which established the interconvertibility between formulae and *terms*.)

The Yellow Book

The logical object which corresponds to the natural numbers, N, in Tractatus, can be interpreted as a *continuously variable set* of natural numbers N related to a *topos construction* satsifying the usual Kock-Lawvere axioms (where the difference between two *external* natural number objects, N_1, N_2 are interpreted as different stages of the construction of the *internal* natural number object N_0). However, rather than just viewing arithmetic as governed by processes of additions m + n of natural numbers m,n, the arithmetic of the Tractatus (WA) is founded on a generalized operation for a *multiple-successor arithmetic*. Thus, given a number m, the value of $m + \ell + 1$, the ℓ+ 1st *successor of* m, is a value which must exist in the arithmetical system WA (given that ℓ exists). On the other hand, the diagonal value m + m of m + n is *undefined* by Tractatus' symbolic conventions [cf. Tractatus 4.0411]. Classically, the diagonalization of the two-place function m + n is denoted by the one-place function 2m (which generalizes to the two-place operation n · m of *multiplication*). The values of the latter function may be used to indicate a formal series of *fixed points* $2 \cdot m, 2 \cdot m + 1, 2 \cdot m + 2, \ldots 2 \cdot m + \ell, \ldots$, at which the arithmetical operation of addition is "flattened out" and overtaken by the arithmetical operations n · m and $\ell + 1$, $\ell,m,n = 0,1,2,3\ldots$. Although the latter constructions of fixed points are not available within the framework of Tractatus, the latter affords other types of fixed point constructions which are related to the growth of the exponents of operations on elementary propositions and their truth-functional combinations.

Now, observe that given any consistent set $\bar{p}$ of elementary propositions ordered with respect to their logical complexity, there are no two elementary propositions p,q, such that q is $\neg p$, i.e. the conjunction $\wedge(p,q)$ is *undefined* whenever q is $\neg p$ or p is $\neg q$. However, if, say, there is an elementary proposition r expressing the fact that a spot in the visual field is *red*, then there could be elementary propositions p,q,..., expressing the fact that the same spot is not-white, not-black,...a series which may be interpreted as giving an increasingly correct "approximation" of the proposition r. That is, if the number of distinct colors were *finite* and, as a set, containing, say, ℓ members, then the ℓ- 1-ary conjunction

$$\bigwedge_{1}^{\ell-1} \{ p, q, \ldots \}$$

of l-1 possibly elementary propositions recursively entails r while their l-1-ary joint denial would be equivalent to $\neg r$, i.e. the elementary proposition p expressing that the spot in the visual field is *not* red. But, obviously, this contradicts the independence of the elementary propositions. An immediate conclusion of this meditation on the logic of colors is that in so far as colors may be comprehended by elementary propositions, they must not be bounded by a finite set, the number of logical colors being numberless (zahllos) i.e. *non-finite*. Using a less archaic language, one says today that the problem of estimating a lower upper bound on the cardinal number of the number of logical colors remains *unsolvable* (even when notations for transfinite ordinals are allowed). And the same goes for membership in $\bar{p}$, i.e. it is *undecidable* whether or not r (or p,q,...) is (are) among the members of the set $\bar{p}$ of elementary propositions p,q,r,...., and, it still remains unknown whether or not the latter sequence is co-extensive with the *empty sequence* of elementary propositions. Yet, if $\bar{p}$ *is* non-empty, then it can be shown that $\bar{p}$'s cardinal number must exceed one, i.e. in terms of (cardinal) size, $\bar{p}$ must be, at a minimum, a 0-*large set* [cf Tractatus 2.0121]

Interpretation of the Above

Quod decet bovem dedecet Jovem.

It is, of course, well-known that the Tractatus text ostensively neglects the more fundamental aspects of the *decision problem [Der Entscheidungsproblem]*[Cf. Tractatus 6.5...: *The riddle* does not exist ...]. *For example*, everybody knows today that it is not even sufficient (given consistency) to identify a formal system, such as WA, with its set of deductions, since one also has to consider with particular detail the *manner in which it is verified that a syntactic object is a deduction*—where, of course, the additional information or "structure" (semantics) being the (formal) sequence involved in building up the deductions themselves. Yet, obviously, there is (assuming Church's thesis) a mechanical method (algorithm) of deciding whether $\underline{d}$ is a (formal) derivation of any *existential formula* $\exists x \Phi x$.

However, this is not the case for *universal formulae*—a topic discussed at length for the first time by Gödel, 1931. Here are some simple examples. Given formulae Φ_1, Φ_2 and Φ_3 of WA one of the following *cannot* be derived in WA:

i. Φ_1 expresses that WA is (deductively) *closed* under *modus ponens* and Φ_1 holds

ii. Φ_2 expresses that WA is *complete* for numerical arithmetic (wrt a specific arithmetical predicate) and Φ_2 holds

iii. Φ_3 expresses that WA is *consistent* and Φ_3 holds

It is rather remarkable that, although not explicitly stated, these facts may be guaged already from Frege's Appendix to *Grundgesetze*, vol.II.

With reference to Φ_3, the concept of *identity* of deductions (proofs) is absolutely critical. *For example*, there are at least two distinct proofs of the formula

(Φ_4) $$p \wedge \neg p \to (p \to p)$$

viz. i. one ignores the conclusion (*ex falso quodlibet*), or, ii. one ignores the premise (since $p \to p$ is a tautology.) Here a formal identification of these corresponding two proofs is obviously a non-trivial problem. Furthermore, in this context it should be stressed that *different styles of formalization* may be compared with the use of different coordinate systems in geometry, where one system is particularly suited for the study of a *specific geometric relation*, "suited" not in the sense of being manageable, but, rather, where the passage from one co-ordinate system to another may introduce singularities *without geometric meaning* (such as the indeterminateness of the second polar co-ordinate at the origin.)

Returning now to Wittgenstein's enigmatic notation

$$(\text{------}W)(\xi,\ldots.) [(\text{------}T)(\xi,\ldots.)]$$

it must be immediately remarked that *what is not* is *not* recursively enumerable (r.e.). That is (even) if *what is* is r.e., its complement is *not* r.e. However, neither is the set of finite models (of this context) r.e. so (a black-and-white) logical space would be *incomplete* for most *finite* ξ. On the other hand, from the incompleteness property one deduces that a sequence of falsehoods of order type N ($= $ "ω^{WA}") will not necessarily correspond to the constant function $\bot$ (designating permanent falsehood) but rather to (by switching non-linearly at infinity) the constant function $\top$ (designating permanent truth) as the following recursion shows (where "$\vdash$" denotes the formally defined *provability operator*):

$\vdash^{0}\bot := \bot$

$\vdash^{k+1}\bot := \vdash(\vdash^{k}\bot)$

$\vdash^{N}\bot := \neg\bot$

Writing ■ for $\bot$ and □ for $\top$ the following diagram relates the *continuum problem* with the *Liar Paradox* (and propositional quantifiers with modal operators).

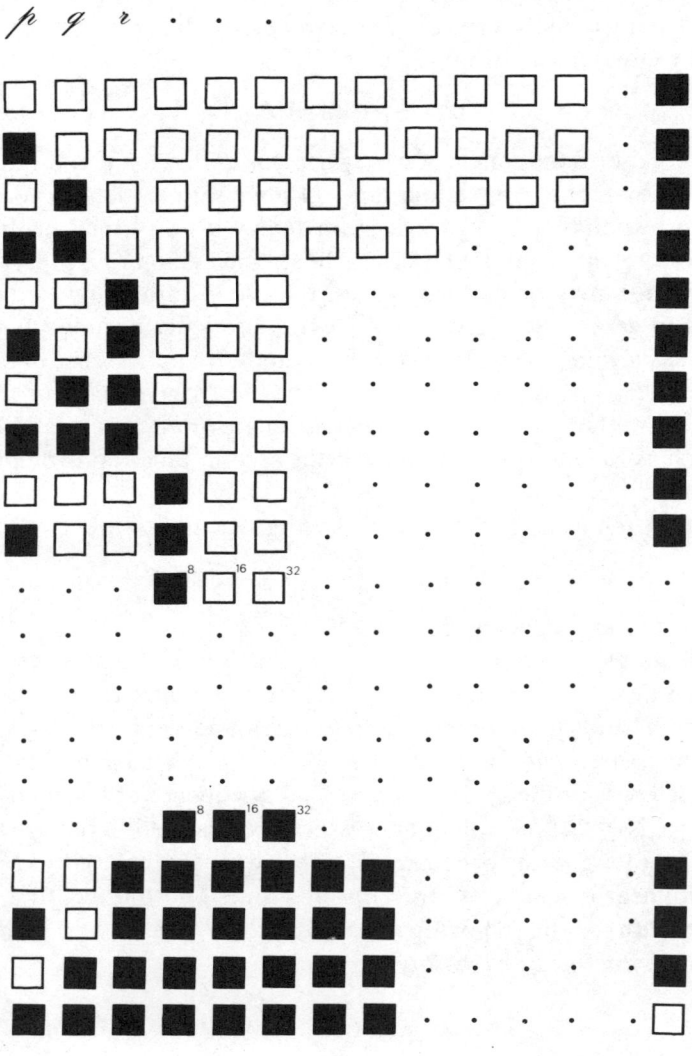

The Yellow Book

A Continuously Variable Open Point Set η of Measure 0

The set η is a *totally disordered* set with no distinguished first or last elements. The elements of η are called *points* and they are indicated (when present) by *black dots* (of variable sizes).

The semiotical action corresponding to the event $e \in \acute{E}$ where the set η is losing some of its points is the *merging* or *identification* of pairs of points $<\beta, \gamma>$ (with a single point α—which *may* also be (non-deterministically) identified with either β or γ (as in))

$$\{\ldots\}_{\alpha\beta\gamma} \mapsto \{\ldots\}_{\alpha\beta\ldots}$$

The dual event where η acquires new points shall be semiotically indicated by the action of *splitting* or *discerning* a single point α, thereby creating a (not necessarily ordered) pair $\{\beta, \gamma\}$ of points, i.e.

$$\alpha \mapsto \{\beta, \gamma\}$$

A more precise notation for the set η is $\acute{E}(\eta)$ together with a *calander signature, e.g* $\acute{E}(\eta)_{41X - 17X\ 76}$.

Different phases of the latter continuously variable set are shown as a *continuously variable floor painting* (reproduced on the following pages). Thus, this floor painting serves simultaneously both as a *picture* and as a *model* for a *finite set of the topoi* constructed in the style of F.W. Lawvere.

Remark. The here documented set $\acute{E}(\eta)_{41X-17X\ 76}$ only existed during the time interval indicated by the calender signature and has since that time never existed again.

Continously Variable Open Point Set Of Measure Zero
(Partial views 4 IX - 17 X 76—Moderna Museet, Stockholm)

Christer Hennix

Epistemological Intermezzo (I,II)

The simplest objects which are directly dealt with are called *"signs"* and among them one distinguishes the *elementary signs* as an enumerable sub-species which intersects any non-empty species of signs.

Let E_0 denote an inhabited species of elementary signs. Then, by $\mathcal{E}_{E_0}$ I indicate a (ω-) *finite* Nn, the image $N_{\mathcal{E}_0}$ of which enumerates $E_0 \cup \{ \triangle \}$. [Here "$\triangle$" is a sign for an occurrence of an empty sign which does not inhabit E_0.]

I shall indicate that a natural number k belongs to $N_{\mathcal{E}_0}$ by the (non-positional) *unary system of indications*

$|, ||, |||, \ldots\ldots, |^k, |^{k+1}, \ldots\ldots$

If ω is the k:th elementary symbol in E_0, then ω denotes the positive integer $k + 1$.

It must be decidable, given any object α, whether $\alpha \epsilon E_0$ and whether $\alpha \neq \omega$. Thus, $N_{\mathcal{E}_0}$ is an intuitively decidable Nn since the map $.: \alpha \to 1^k$ is required to be effective.

The decidability of the species of elementary signs is a necessary property for any *direct* correctness proofs [proofs of absence of errors]. On the other hand, the sequences of species of non-elementary signs is not even r.e. and neither are many of its members $E_1, E_2, E_3, \ldots\ldots, E_k, E_{k+1}, \ldots\ldots$ In particular, its empty members, E_ϕ, are *not* r.e.

For each inhabited species E_j, there shall be an Nn, $N_{\mathcal{E}_j}$, which depends both on E_j and the function $\mathcal{E}_j$ enumerating E_j.

Thus, the continuum of inhabited species of non-elementary signs shall be denoted by $\underset{\cdot}{E}$—and its Nn by $N_{\underset{\cdot}{E}}$.

Reminder.

Correct sequences of elements of $\underset{\cdot}{E}$—such as *programs*—are among the most complex ontological objects so far encountered from searches at this end of the universe (Poets must ask themselves (again) if Mallarmé was not right—after all.—And, on reflection, must not proof-theorists, if not novelists, take Robbe-Grillet's dictum, "Nothing is more fantastic, ultimately, than precision" into renewed considerations?!)

SPECTRA OF MODALITIES
AND
THE THEORY OF THE CREATIVE SUBJECT, $Th(\Sigma)$

I. It is a well-known fact that activities ending in art may not always carry a well-determined meaning or sense. On the contrary, the lack of meaning(-fulness) is compensated for by (purported) aim(s) expressed by the purpose(s) governing the installment of the generating activities for which end some particular object is taken as a witness. This somewhat confused reality shows the indisputable importance of the interpretation of the <u>optative modalities</u> underlying any goal-oriented activity $\mathcal{A}$.

II. In order to fix the purposefulness of an activity $\mathcal{A}$, several ratios are to be measured, such as

1) $$\frac{\text{Interest of results}}{\text{Efforts involved}}$$

AND

2) $$\frac{\text{Long-term satisfaction}}{\text{Efforts required}}$$

Given some satisfactory measures of these ratios for an activity $\mathcal{A}$ there is a further requirement on the <u>means</u> available by which the result(s) of $\mathcal{A}$ are achieved. Viz. it is generally required that any aim achieved through $\mathcal{A}$ is acquired only as far as <u>fair</u> means have been provided.

323

Thinking occurs as a form in many variables.

Forms of thinking are forms of attention. (No discernment takes place if attention is not present.) Acts of attention preceed every identification.

(When the Creative Subject discerns or identifies the values of her current free variables she is already enduring a succession of acts of attention. Here shall only be mentioned attention to variables and attention to forms.)

(The geometry of forms is a notation for the projective content of thought, i.e., the points of thought, when charted on a manifold, that are determined by the values of the free variables of its form(s).)

When the Creative Subject identifies two thoughts or thought processes by means of yet another thought the two thoughts (or thought processes) are said to be of the *same type*.

(Types of thought are discerned by their (intrinsic) evaluation of the Creative Subject's current free variables.

From a geometric point of view thought has the shape of a manifold along which certain variables have been evaluated.)

The form of thought which expresses attention to form is *also* a form of attention. In the limit, this form can be expanded as a *manifold of attentions* (mastered by the Creative Subject).

Attention penetrates thought (in) as much as thought penetrates attention. Attention combined with thought is an operator of (intensional) discernment, a form of a complex perception in a space where mind and cosmos momentarily coalesce along a singular manifold of arbitrarily many dimensions.

In this way, existence becomes, ultimately, a result of attention (or absence of attention). And attention is the result of a form of thought. What is is thus firmly trapped in our forms of thought. (or forms of absence of thought).

Grammatica tua sit tibi in perditionem!

Any violation of this <u>fairness principle</u> is to be considered harmful for the continuation of the situations S generated by $\mathcal{A}$, on account of the <u>displacement of modalities</u>, caused, in particular, by the displacement of goals relative the installment of $\mathcal{A}$.

III. Clearly, the condition of fairness for purposeful activities $\mathcal{A}$ imposes an obvious restriction as to the availability of means for realizing $\mathcal{A}$. This restriction must be evaluated relative the higher-order aims under which $\mathcal{A}$ is subsumed. As far as clarity and certainty is sought, the fairness principle is but a higher-order means for our epistemic development, and its violation poses obstacles as far as (foundational) communication is aimed at (to wit, its deepest threat).

IV. For the purpose of <u>restricting</u> the above-mentioned restriction, two <u>spectra of modalities</u> are defined for the sake of optimal freedom in activities $\mathcal{A}$ restricted by fair means. Viz.

(1) The first spectrum which will be designated Freedom$_1$ or F$_1$ and is defined as that <u>regime</u> $\mathbf{P}$ governing activities $\mathcal{A}$ in the <u>absence of any obstructions</u>. Id est, F$_1$ assigns the following interpretation to the fulfilment of the optative modalities connected with $\mathcal{A}$: If T is an aim in $\mathcal{A}$ and $\mathcal{A}$ provides (α) sufficient means and (β) all necessary means for realizing T in $\mathcal{A}$, then T is fulfill<u>able</u> in $\mathcal{A}$. F$_1$ clearly corresponds to the <u>purposefulness</u> of $\mathcal{A}$ and would be violated in any situation where (α) (β) hold but T has been (or will be) obstructed.

Christer Hennix

Main Subject.
Re: Tractatus 3.144 [Names are like points, propositions like arrows].

The world consists of arrows. The
                        world disintegrates
                        in arrows.
Each arrow is a fact—a mental fact
                    a mutual fact.
Each fact is an arrow.

Arrows run between points. The leftmost point serves as a comprehension term for the rightmost point.

(A philosophical *problem* can sometimes be analyzed as a *stack of arrows*, in which case its *solution* can be given as—a bundle of paths going through a *forest of arrows*.)

The simplest arrow is the

empty arrow

it designates a point of *negative dimension* ≤ -1.

Clearly, the empty arrow does not consititue a philosophical problem—unless the negative dimension numbers do. *Via negativa*, again, nor does it constitute a philosophical solution—unless, that is, it points to negative infinitesimal dimension numbers close to, but not identical with, the points of $-\infty$, the darkest dimension of cosmic existence.

Presently, only

■   designates the *continuum of empty arrows*

[points here play the role of comprehension terms for collections of empty spaces]

☐   designates the *unit space* of a constant arrow

$$\alpha \to \beta$$

(2) The second spectrum which will be designated Freedom$_2$ or F$_2$ and which is defined as that <u>regime</u> **P** governing activities $\mathcal{A}$ such that <u>no act in</u> $\mathcal{A}$ <u>is forced by coercion, fraud or any other violation of the fairness of means provided for the course of</u> $\mathcal{A}$. Id est, $\mathcal{A}$ is said to possess property F$_2$ whenever every act in $\mathcal{A}$ is free from compulsion, id est exercised in accordance with the creative subject's free will. Clearly, the satisfiability of F$_2$ captures exactly what is intended by <u>justfulness</u> of $\mathcal{A}$ and the <u>spectrum of</u> F$_2$-modalities is precisely all instances of installments of given activities $\mathcal{A}$ for which fair means have been provided.

V. The spectra of modalities for $\mathcal{A}$ thus split into two basic components, F$_1$ and F$_2$. By passing to the <u>direct limit</u> of the projections of F$_1$ and F$_2$ on any $\mathcal{A}$, the composite F$_1$ F$_2$ obtains. It corresponds to that <u>regime</u> **P** for which Freedom$_1$ and Freedom$_2$ hold <u>simultaneously</u> and if $\mathcal{A}$ possesses property F$_1$ F$_2$ we shall say that **P** is <u>eleutheric</u> for $\mathcal{A}$ or that the activities comprised by the situations generated by $\mathcal{A}$ are <u>eleutheric activities</u> in the regime **P**

<u>Axioms for</u> Σ :

I: $\Sigma \vdash_n \mathscr{A} \; \mathbf{v} \; \Sigma \vdash_n \neg \mathscr{A}$

II: $\Sigma \vdash_n \mathscr{A} \rightarrow \mathscr{A}$

III: $\Sigma \vdash_n \mathscr{A} \; \& \; m > n \; \rightarrow \; \Sigma \vdash_m \mathscr{A}$

IV: $\neg \exists (x) \; \Sigma \vdash_x \mathscr{A} \rightarrow \neg \mathscr{A}$

A unit space of dimension 1 is obtained by evaluating the arrows

$$\{\Lambda\} \to \Lambda$$

as

$$\Lambda \in \{\Lambda\}$$

Reflection on negative dimension numbers < -1 yields the higher dimensional unit spaces $\square_\alpha, \alpha > 1)$.

Each sentence is an arrow.

—equations are all arrows.

A sentence is *true* if it points to the higher dimensional arrows—if its associated system of true equations are all arrows.

Otherwise, it is *false* or *absurd*.

Hence, it is absurd that all sentences are false. But it is not true that there is a sentence which is *not* absurd!

Whence, we continue our sacrifice.

Model Subject Φ (Φ)
Frege's 'The Real Way Is Not-Not Difficult'
(Don't write what you write, *or*, write *nothing*)

$$\nu \left[\begin{array}{l} \mathfrak{g} \left[\begin{array}{l} \mathfrak{g}(a) \\ M_\beta(\text{---}\mathfrak{g}(\beta)) = a \end{array} \right. \\ M_\beta\left(\text{---}\mathfrak{g}\left[\begin{array}{l}\mathfrak{g}(\beta)\\ M_\beta(\text{---}\mathfrak{g}(\beta))=\beta\end{array}\right.\right) = a \end{array}\right.$$

(IIIa): ─────────────────────

$$\left[\begin{array}{l} f(a) \\ M_\beta\left(\text{---}\mathfrak{g}\left[\begin{array}{l}\mathfrak{g}(\beta)\\ M_\beta(\text{---}\mathfrak{g}(\beta))=\beta\end{array}\right.\right) = a \\ f(a) = \text{---}\mathfrak{g}\left[\begin{array}{l}\mathfrak{g}(a)\\ M_\beta(\text{---}\mathfrak{g}(\beta))=a\end{array}\right. \end{array}\right.$$

(ξ

(IIb):: ─ ─ ─ ─ ─ ─ ─ ─ ─ ─ ─ ─

$$\left[\begin{array}{l} f(a) \\ M_\beta\left(\text{---}\mathfrak{g}\left[\begin{array}{l}\mathfrak{g}(\beta)\\ M_\beta(\text{---}\mathfrak{g}(\beta))=\beta\end{array}\right.\right) = a \\ M_\beta(\text{---}f(\beta)) = M_\beta\left(\text{---}\mathfrak{g}\left[\begin{array}{l}\mathfrak{g}(\beta)\\ M_\beta(\text{---}\mathfrak{g}(\beta))=\beta\end{array}\right.\right) \\ \mathfrak{F}\left[\begin{array}{l}\mathfrak{F}(a) = \text{---}\mathfrak{g}\left[\begin{array}{l}\mathfrak{g}(a)\\ M_\beta(\text{---}\mathfrak{g}(\beta))=a\end{array}\right.\\ M_\beta(\text{---}\mathfrak{F}(\beta)) = M_\beta\left(\text{---}\mathfrak{g}\left[\begin{array}{l}\mathfrak{g}(\beta)\\ M_\beta(\text{---}\mathfrak{g}(\beta))=\beta\end{array}\right.\right)\end{array}\right.\end{array}\right.$$

(o

INFERENCE SCHEMA FOR $T_H(\Sigma)$

$$V: \frac{\Sigma \vdash_n \mathscr{A}, \quad \mathscr{A} \to \mathscr{C}}{\Sigma \vdash_n \mathscr{C}}$$

$$VI: \frac{\Sigma \vdash_n \exists(x) \mathscr{A}(x)}{\exists(x) \Sigma \vdash_n \mathscr{A}(x)}$$

$$VII: \frac{\Sigma \vdash_n \mathscr{F}, \quad \Sigma \vdash_m \mathscr{F} \to \mathscr{G}}{\Sigma \vdash_{n,m} \mathscr{G}}$$

The general idea is now as follows. The basic relation $\vdash$ in the context $\Sigma \vdash_n \mathscr{A}$ is interpreted as "Σ has <u>decided</u> or <u>solved</u> $\mathscr{A}$ at the n^{TH} stage of her investigation or research in $\mathcal{A}$", where $\mathscr{A}$ designates an <u>intension</u> or <u>problem</u> connected with $\mathcal{A}$. The logical operators $\neg$, $\&$, $\vee$ and $\to$ are given the following interpretations.

$\neg[\mathscr{A}]$ signifies the task of obtaining the absurdity of the solution of $\mathscr{A}$

$\&[\mathscr{A}, \mathscr{B}]$ " " " " solving <u>both</u> $\mathscr{A}$ and $\mathscr{B}$

$\vee[\mathscr{A}, \mathscr{B}]$ " " " " " <u>either</u> $\mathscr{A}$ or $\mathscr{B}$

$\to[\mathscr{A}, \mathscr{B}]$ " " " " " the problem $\mathscr{B}$ given any solution of $\mathscr{A}$

(IIb, IIIa) :: = = = = = = = = = = = = = = =

$$\vdash\begin{array}{l}f(a)\\ M_\beta(\text{---}f(\beta))=a\\ M_\beta\left(\begin{array}{l}\mathfrak{g}\\ \text{---}\\ \end{array}\begin{array}{l}\mathfrak{g}(\beta)\\ M_\beta(\text{---}\mathfrak{g}(\beta))=\beta\end{array}\right)=a\\ \mathfrak{G}\ \mathfrak{F}\ \mathfrak{F}(a)=\mathfrak{G}(a)\\ M_\beta(\text{---}\mathfrak{F}(\beta))=M_\beta(\text{---}\mathfrak{G}(\beta))\end{array}$$ (π

$$\vdash\begin{array}{l}\mathfrak{g}(a)\\ M_\beta(\text{---}\mathfrak{g}(\beta))=a\\ M_\beta\left(\begin{array}{l}\mathfrak{g}\\ \text{---}\\ \end{array}\begin{array}{l}\mathfrak{g}(\beta)\\ M_\beta(\text{---}\mathfrak{g}(\beta))=\beta\end{array}\right)=a\\ \mathfrak{G}\ \mathfrak{F}\ \mathfrak{F}(a)=\mathfrak{G}(a)\\ M_\beta(\text{---}\mathfrak{F}(\beta))=M_\beta(\text{---}\mathfrak{G}(\beta))\end{array}$$ (ϱ

(μ) : - - - - - - - - - - - - - - -

$$\vdash\begin{array}{l}M_\beta\left(\begin{array}{l}\mathfrak{g}(\beta)\\ M_\beta(\text{---}\mathfrak{g}(\beta))=\beta\end{array}\right)=a\\ M_\beta\left(\begin{array}{l}\mathfrak{g}(\beta)\\ M_\beta(\text{---}\mathfrak{g}(\beta))=\beta\end{array}\right)=a\\ \mathfrak{G}\ \mathfrak{F}\ \mathfrak{F}(a)=\mathfrak{G}(a)\\ M_\beta(\text{---}\mathfrak{F}(\beta))=M_\beta(\text{---}\mathfrak{G}(\beta))\end{array}$$ (σ

(Ig) : - - - - - - - - - - - - - - -

$$\vdash\begin{array}{l}M_\beta\left(\begin{array}{l}\mathfrak{g}(\beta)\\ M_\beta(\text{---}\mathfrak{g}(\beta))=\beta\end{array}\right)=a\\ \mathfrak{G}\ \mathfrak{F}\ \mathfrak{F}(a)=\mathfrak{G}(a)\\ M_\beta(\text{---}\mathfrak{F}(\beta))=M_\beta(\text{---}\mathfrak{G}(\beta))\end{array}$$ (τ

(IIa) :: - - - - - - - - - - - - - - -

$$\vdash\begin{array}{l}M_\beta\left(\begin{array}{l}\mathfrak{g}(\beta)\\ M_\beta(\text{---}\mathfrak{g}(\beta))=\beta\end{array}\right)=a\\ \mathfrak{a}\ \mathfrak{G}\ \mathfrak{F}\ \mathfrak{F}(\mathfrak{a})=\mathfrak{G}(\mathfrak{a})\\ M_\beta(\text{---}\mathfrak{F}(\beta))=M_\beta(\text{---}\mathfrak{G}(\beta))\end{array}$$ (υ

$\times$

$$\vdash\begin{array}{l}\mathfrak{a}\ \mathfrak{G}\ \mathfrak{F}\ \mathfrak{F}(\mathfrak{a})=\mathfrak{G}(\mathfrak{a})\\ M_\beta(\text{---}\mathfrak{F}(\beta))=M_\beta(\text{---}\mathfrak{G}(\beta))\\ M_\beta\left(\begin{array}{l}\mathfrak{g}(\beta)\\ M_\beta(\text{---}\mathfrak{g}(\beta))=\beta\end{array}\right)=a\end{array}$$ (φ

———•———

IIIe $\vdash M_\beta\left(\begin{array}{l}\mathfrak{g}(\beta)\\ M_\beta(\text{---}\mathfrak{g}(\beta))=\beta\end{array}\right)=M_\beta\left(\begin{array}{l}\mathfrak{g}(\beta)\\ M_\beta(\text{---}\mathfrak{g}(\beta))=\beta\end{array}\right)$

(φ) : ————————————————————————

$$\vdash\begin{array}{l}\mathfrak{a}\ \mathfrak{G}\ \mathfrak{F}\ \mathfrak{F}(\mathfrak{a})=\mathfrak{G}(\mathfrak{a})\\ M_\beta(\text{---}\mathfrak{F}(\beta))=M_\beta(\text{---}\mathfrak{G}(\beta))\end{array}$$ (χ

IN ADDITION:

$\exists(x)[\mathcal{A}(x)]$ SIGNIFIES THE TASK OF SOLVING THE VALUE OF x SUCH THAT $\mathcal{A}(x)$

WHERE $\exists(x)$ IS THE OPERATOR DESIGNATING THE <u>EXISTENTIAL CONSTRUCTION PRINCIPLE</u>.

AMONG SOLVABLE PROBLEMS IN $\text{Th}(\Sigma)$ WE MENTION THE FOLLOWING TASKS (EXERCISE!):

$$\mathcal{A} \to \mathcal{B} \,\&\, \mathcal{B} \to \mathcal{C} \to \mathcal{A} \to \mathcal{C} \qquad (1)$$

$$\mathcal{A} \to \mathcal{C} \,\&\, \mathcal{B} \to \mathcal{C} \to \mathcal{A} \vee \mathcal{B} \to \mathcal{C} \qquad (2)$$

$$\neg \mathcal{A} \to \mathcal{A} \to \mathcal{B} \qquad (3)$$

$$\mathcal{A} \to \mathcal{B} \to \neg \mathcal{B} \to \neg \mathcal{A} \qquad (4)$$

$$\mathcal{A} \to \neg \mathcal{B} \to \mathcal{B} \to \neg \mathcal{A} \qquad (5)$$

$$\neg(\mathcal{A} \vee \mathcal{C}) \to \neg \mathcal{A} \,\&\, \neg \mathcal{C} \qquad (6)$$

$$\neg(\mathcal{C} \,\&\, \mathcal{E}) \,\&\, \mathcal{C} \vee \neg \mathcal{C} \to \neg \mathcal{C} \vee \neg \mathcal{E} \qquad (7)$$

$$\neg(\mathcal{A} \,\&\, \neg \mathcal{A}) \qquad (8)$$

<u>REMARK</u>: BY (8), <u>CONSISTENCY</u> OF $\text{Th}(\Sigma)$ IS ESTABLISHED, I.E. $\text{Th}(\Sigma)$ IS A <u>NON-TRIVIAL</u> THEORY (I.E. EPISTEMICALLY PALATABLE).

<u>(HISTORICAL) REMARK</u>: <u>L.E.J. BROUWER</u> WAS THE FIRST LOGICIAN TO FORMULATE PRINCIPLES SIMILAR TO THE ONES GIVEN ABOVE.

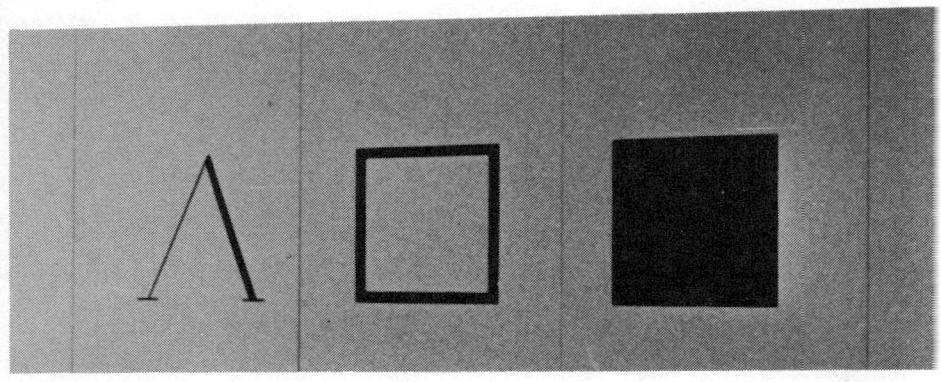

Short Infinitary Process 4 IX—17 X 76
(Moderna Museet, Stockholm)

Idios Kosmos (I)

Pre-socratic Ontologies

Ṛta—concept of *correct action*.
Ȧncient composition of arrows. (Laws.)

0. Only the empty arrow remains the empty arrow when the empty arrow is composed with the empty arrow.
   –When any other arrow is composed with the empty arrow, it remains itself, whether or not the composition takes place to the left or to the right of the empty arrow.

Event $\breve{E} = \acute{E}_{4IX-17X76}$ Def.
$\underset{.}{R}: \breve{E} \to \acute{E}$ $\qquad$ ⸰ (Being $=$ Space $\times$ Action)

Ultra-Environments $\breve{E}$

Types of Universes

(I) .... A, B, Γ, ...
(II) .... a, b, c, d, e′, e″, ...
(III) .... I, II, ...

I. **Monistic Universes** A.

Ia. Σ contemplates the existence of *empty universes*

$$\Lambda\Lambda\Lambda\ldots\ldots\Lambda\Lambda\ldots\ldots$$

and the embeddings

$$\underset{\cdot}{R}: E^\Lambda \to \acute{E}$$

The event Λ may be completely indeterminate —*or*, it may correspond to a designated event which occurs interior to the event $\acute{E}$. *For example*, the event Λ is the genesis of an *empty event* in the interior of $\acute{E}$ at a time τ, *or*, Λ is the genesis of a progression of clusters of *empty cosmoi* experienced by Σ in the interior of $\acute{E}$ at any time τ. The identity of the interior event E^Λ shall be determined by the *inverse image* of the functorial action $\underset{\cdot}{R}$ of the realization map

$$E^\Lambda \to \acute{E}$$

Remark that it is quite possible that $R^{-1}(\acute{E}) \neq E^\Lambda$ because of lack of certain symmetries (such as commutative diagrams!).

Various Notions of Emptiness (Concepts of Absence)
(Partial summary)

> The Empty Cosmic Cluster
> The Empty Arrow
> The Empty Self
> The Empty Thought
> The Empty Text
> The Empty Action/Event

Various Notions of E^Λ (Partial summary)

> —Empty Mā
> —Empty Nō
> —Empty Butō
> —Empty Space-Time

The Yellow Book

MĀ-Ontologies. Case 1, $\bar{X} = 0$).

$M\bar{A} = MA_{\bar{X}}$ (Mirror Duality)

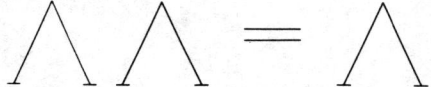

B/W Equation

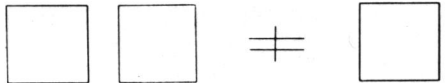

B/W Equation

Monistic Universes B

Ib. Σ contemplates the existence of *non-empty universes*

$$\square\square\square \ldots \ldots \square\square \ldots \ldots$$

and the embedding(s)

$$R: E^\square \to \acute{E}$$

Again, the identity of a non-empty monistic universe □ shall be determined by an inverse limit or image $R^{-1}(\acute{E})$. And, again, it is possible that $R^{-1}(\acute{E}) \neq R(E^\square)$ (if the reality of the situation is lacking in symmetries.)

Standard Examples (Classical Set Theory)

Singleton Sets: {·}, {Λ}, {e}, . . . &c.

É-Examples: Soliton Compositions. (Solitones).

0. *Agrapha Harmonia.*
[*Cognitio Matutina*]

0.1 　　　　The Sound of Shiva—ÔMSAHASTRANAMAM
[For Pandit Pran Nath]

0.11　$\square_\kappa$　　—The κ-times Repeated Constant Event.
[Infinitary Composition. Locally Infinite Space.
Fix Point Structure, $\kappa = \kappa + 1$].

0.111　$\square^N$　　The N-times Repeated Constant Event.
[The Five Times Repeated Music, &c.]

0.111a　　　$N = 3$. Short Infinitary Processes. *White* M̄A.

$$\square \; \square \; \square \Rightarrow . \square \ldots$$

[where "." refers to the empty space left by the left □ merging with its successor and ". . ." to the transfinite term which covers the future of the constantly present boundary or middle term □. (Cf. the Black & White Algebra of the formal (binary) ontology Λ □ ■).]

The Yellow Book

0.1111   *The N-Times Repeated Constant Event* [Version For Sound Waves]

0.1111a   **N = 0. 0-D** Toplogies For A Vertical Sound. The *Zero-Times Repeated Event*, $\square^0$, or *The Sound of MA* is defined by the *Sound of an Indestructible Moment of Silence* contained by a space equipped with a vertical topology.

$\square^0$ —The Empty Sound

0.1111b. $\square_{<x,y,z>}$ *Crystal Oscillators* (Quartz). *Digital Wave Forms.* (Programs)

Transfinite Case;

$\square_\kappa$ —The κ-Infinite Sound

$\square_{<x,y,z>}$—x,y,z transfinite real variables.

The complexity of *constant events* may be measured in various ways. A convenient complexity measure for a constant event e shall be a certain ordinal number ord(e). The non-empty constant events

$$E^{\square}{}_\alpha \qquad \square\square\square \ldots \square\square \ldots \square$$

are usually characterized by *limit ordinals* α. Ord ($\square$) and ord ($\blacksquare$) are two distinct species of limit ordinals which define two families of initial segments of the continuum of ordinal numbers, Ω, and which (naturally) give rise to two (families of) *natural number series* $N_\square$ and $N_\blacksquare$ respectively. The series of constant (canonical) empty events $\square^0$ is usually denoted N_Λ [M$\bar{\text{A}}$-ordinals].

(Summary). The constant events $\square^N$ are defined by the (point-wise) composition of the three distinct curves

$\square_{<x>} : a \times \sin(x)$

$\square_{<y>} : b \times \sin(y)$

$\square_{<z>} : c \times \sin(z)$

the continuous point-wise compositions of which are omnidirectionally distributed in the interior of É as a mixture of a triple of phase-locked sine waves at constant frequencies and amplitudes such that the objects $\square^N$ or $\square_\kappa$ obtain as the respective limit constructions of this continuous process of composition based on fixed (eternal) quartz crystal oscillations in pre-determined ratios.

More specifically, each sine-wave compactum $\square_{<w>}$ describes an infinitely proceeding sequence of constant events $\square_{<w_i>}$ the fixed amplitude of which is inversely proportional to the frequency of the constant events that build up this indefinitely on-going wave structure. The possible frequencies of these constant events are determined by a certain set of *odd* prime numbers. Thus, 2 certainly does *not* belong to this set, but, say 5 *may* belong to it; or all prime numbers larger than any large but otherwise arbitrary prime number may belong to it with certainty. Hence this set may certainly be infinite. If the kth prime p_k does not belong to this (possibly infinite) set and p_k divides m/n, the ratio between two sine-wave components, then any ratio m:n:p_k is a possible (but not necessary) determination of a sine-wave component in a (É-)realization of the composition $\square^N$ or its transfinite extension $\square_\kappa$. (The eternal presence being constantly controled or determined by what is eternally absent.)

Elements which belong to the complex wave form $\circledast^N$ are called *chronons* or *time elements*. Here the technique of *analytic continuation* is presupposed as well as the most efficient programs for the **FFT** [Fast Fourier Transform.]

At the moment when the Creative Subject enters a frame for a topos $\mathcal{E}$ as defined by the interior of É, the time manifold $\square^N$ is supposed to already contain indefinitely many time elements which have not yet or never will be experienced by the Creative Subject.

The Yellow Book

The moment at which the Creative Subject's tactics of attention include attention to the time process ⊞^{N}, corresponds to a point in her life-world where a moment of life falls apart with one part retained as an image and stored by memory while the other part is retained as a continuum of new perceptions.

This marks the beginning phase of the Creative Subject's experience of the *fixed point composition* $\Box^N$, i.e.

The N-times Repeated Constant Event

[begun by me during the winter months of 1970 and continued into the present].

In a later phase of the Creative Subject's experience of $\Box^N$, the obtaining continuum of perceptions becomes gradually attenuated by a *continuum of memories* of a N-times repeated constant event ⊡. Thus, the Creative Subject gradually comes to experience $\Box^N$ as a *transfinite object* order-isomorphic to the set of rational numbers η (where the latter may be identified as *pairs* of integers m,n, except for the rational number 0 (zero) which may be identified with the usual 0 in the set of natural numbers.)

In particular, at this later stage of experiencing $\Box^N$ the Creative Subject does experience the conjunction of the continuum of perceptions and the continuum of memories as a *dense ordering of (the falling apart of) moments of life,* i.e., an ordering without first or last elements and of Cantorian order-type η.

Finally, at the "transfinite stages" of experiencing $\Box^N$, the Creative Subject arrives at a stationary subset of the generated continuum of perceptions at which she retains complete facultative control of the continuum of time-elements which defines the transfinite time-object ⊞$_\kappa$, [here κ denotes (an ordinal of) some "large cardinal" and not merely a "large number" as connoted by the symbol N].

Digression

I notice in passing that the formation of the pairs

$\langle \Sigma, \Box^N \rangle$ (◈—Σ)

and

$\langle \Sigma, \Box_\kappa \rangle$ (◈—Σ)   (in particular)

gives rise to the formation of (subjectively experienced) *locally infinite spaces* (whose elements consist of correlations between moments of consciousness and time-elements belonging to the continuum of perceptions of the invisible time-manifolds $\square_\kappa$).

In particular, the composition $\square_{<x,y,z>}$ acts as a *covering* Cov(**F**) of a frame **F** of a topos $\mathcal{E}$. That is, the lattice of infinitesimal constant time-manifolds $\square$ *has a dense embedding in the interior of the topos frame if the latter is understood as already containing the finer (transfinite) lattice* $<\square_\kappa>$ obtained by projecting $\boxdot_\kappa$ chaotically at each lattice point.

$$\square^{N_m} \xmapsto{N} \square^N$$

For m = 1,2,3,4,5..., $\square^{N_m}$ is a *localization* of $\square^N$ in a topos frame **F**.

$$\bigcup_{k=1}^{m} \square^{N_k} \;\middle|\; = \begin{array}{l} \text{The Composite Sound Wave } \square_{<x,y,z>} \text{ Covering } \mathbf{F} \\ \text{The Infinitesimal Time-manifold } \mathbf{F}(\tau) \text{ covering } \mathbf{F} \text{ (at} \\ \text{any instantaneous moment } \tau) \end{array}$$

Each localization $\square^{N_k}$ is a terminal object with a chaotic topology. (Single arrow)

The covering space Cov(**F**) is a totally disconnected topological space of neighborhoods with a chaotic topology.

2-D detail of the lattice Cov(**F**)

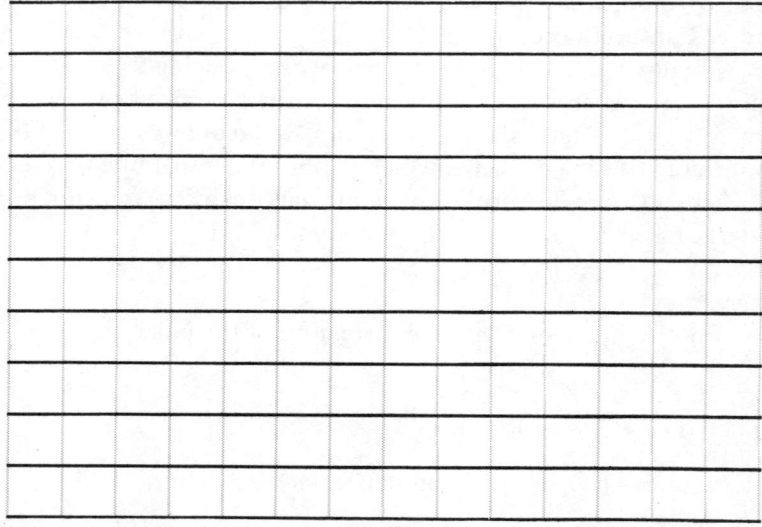

The Yellow Book

The disconnected space of subobjects of the time-manifold $\square^N$ covering the interior of **F** restricts both horizontal and vertical movements by the attending Creative Subjects. Each attending Creative Subject may only compose with movements of infinitesimal arc-lengths, so that stasis dominates over movement and movement, when manifest, occurs only at infinitesimal velocities.

[*(Ultra-) Black Metastasis:* Framed Abstract Butō [Sub-spaces of a Black Topology. See appendix II.]

(Summary cont.)

There exists a *unique time-interval*, τ, in the space **É** defined by the length of the period of the resulting *composite sound wave form* $\square_{<x,y,z>}$. When the time-interval τ has repeated itself **N** times, the time-arc τ_N corresponds to the repetition of the constant event $\square$ **N** times—also written $\square^N$, or, more accurately,

$$\square^N_{<x,y,z>}$$

Remark that the unicity of the infinitessimal time-arc element τ is here based squarely on *geometrical* considerations so that the *constancy* of the invisible event $\square^N$ may be defined in terms of a unique geometrical congruence between the areas of the curves (boundaries)

$$\square = \square^1, \square^2, \square^3, \ldots, \square^k, \square^{k+1}, \ldots, \square^N = \square^{N+1}$$

which, in turn, are defined by a point-wise composition of images under the analytic functions $\lambda \xi \sin(\xi)$.

The triad $<x,y,z>$ defines a class of equations $E(x,y,z)$ which in turn determines a certain set of *algebraic theories*, Th(**E**), among which *triples* $<X,Y,Z>$ and co-triples $<X^{co}, Y^{co}, Z^{co}>$ have played an important role in my early formulations of an ALGEBRAIC AESTHETICS

For each Creative Subject Σ who attends the space $\text{É}_\square$ in which the composition $\square^N$ is freely developing, there exists a (subjective) time-interval, τ^*, during which the Creative Subject experiences a *constant event* $\square^*_{<x,y,z>}$ as a *subobject* of the $\text{É}_\square$—universal infinitely proceeding sequence of geometrically congruent composite wave forms $\square^\tau_{<x,y,z>}$.

In general, the constancy of the subjective event $\square^*_{<x,y,z>}$ can be experienced as also fluctuating so that the subobject $\square^*_{<x,y,z>}$ of $\text{É}_\square$ (created by a Creative Subject's attendance) as a submanifold $\square^{N*}$ of $\text{É}_\square$ defines a *continuously or discretely variable set* (of (fluctuating) experiences of a temporal constancy of time).

Thus $\square^*_{<x,y,z>}$ corresponds to Σ's perception or image of $\square_{<x,y,z>}$ along the subjective time-arcs τ^* iterated N^*times. Obviously, $N^* < N$.

In consequence of the latter inequality, the N^*-times iterated time-element τ^* may be made to correspond to an "arrow of time" in $\acute{E}_\square$ (point-wise speaking).

Thus, $\acute{E}_\square$ is obtained as a subjective space-time manifold of indefinitely iterated invisible constant events $\square_{<x,y,z>}$.

$\acute{E}^\nu$ $\qquad \square_0 = 1, \square, \square_2, \square_3, \ldots ,$

are considered the objects of this category and consequently

$$1 \to \square, \square \to \square_2, \ldots ,$$

are its morphisms.

More precisely, $\square^N$ has as *objects*

$$1, \square, \square_2, \square_3, \ldots ,$$

and for each $n = 0, 1, 2, 3, \ldots, n$ *morphisms*

$$\square^{N_n} \xrightarrow{\Pi_i^{(n)}} \square, \qquad i = 0, 1, \ldots, n-1,$$

such that for any n morphisms (in $\square^N$)

$$\square^{N_n} \xrightarrow{\theta_i} \square, \qquad i = 0, 1, \ldots, n-1,$$

there exists exactly one morphism

$$\square^{N_n} \xrightarrow{<\theta_0, \theta_1, \theta_2, \ldots, \theta_{n-1}>} \square^{N_n},$$

so that, in $\square^N$,

$$<\theta_0, \theta_1, \theta_2, \ldots, \theta_{n-1}> \Pi_i^{(n)} = \theta_i, , i = 0, 1, \ldots, n-1.$$

If the arbitrary morphisms

$$\square^{N_m} \xrightarrow{\phi} \square$$

are regarded as n-ary operations of $\square^N$, then $\square^N$ can be regarded as the *category of constantly shifting events* (which is simultaneously an *algebraic theory*), beginning with the event $1 \to \square$.

B/W Equation

Christer Hennix

Monistic Universes Γ. *(Limit Cases. Ultra-monism)*

1c. Analogously, Σ will contemplate the existence of *large monistic universes*

■■■ ...... ■■ ......

and the imbedding(s)

$$R: E^\bullet \to \acute{E}.$$

Remark. Already the figure "2" is here considered to denote a *"large number"* in the sense that "2" corresponds to the *smallest* large number (among all numbers, 2 is "0-*large*".) Thus, 1 is the largest "small number" and 0 is the smallest such number (0 is "0-*small*".)

Consequently, the set of *small numbers*

(S) $\quad\quad\quad\quad\quad\quad$ {0, 1}

is a (0-large) finite set and it is unique if "0" and "1" are unique. The complement of S is

(L) $\quad\quad\quad\quad$ {2, 3, 4, ..., $\ell, \ell+1, ...$},

the *set of large numbers*—and it is usually *not* a unique set, unless "$\ell+1$" is unique. Furthermore, this set can be *very large* (in comparison with S).

The set of natural numbers, N, is sometimes partitioned into small and large numbers. By

□

I shall indicate a *generic small set* of natural numbers, while

■

shall denote a *generic large set* of natural numbers. In particular, for any generic brand N of natural numbers I may write

$$N = \square \cup \blacksquare, \text{ or,}$$

$$N = \cup(\square, \blacksquare)$$

to indicate that N is a *generic set (brand) of natural numbers*.

[The well-known inductive paradox by which it is shown that all natural numbers are "small" can be resolved already by this generic partition of N. That is, the critical inductive step at "if n is a small number, then n + 1 is also a small number," becomes vacuous (or "generic") by the identification n = ℓ since it entails that $\ell \leq 1$, 1 being considered the *largest small number*.

The Heap Paradox. This paradox is a dual variant of the last one in the sense that closure under n ± 1 (predecessor arithmetic) appears to fail. Hence, the appropriate induction principle sought is Brouwer's principle of *Bar Induction*, i.e.

$$\frac{\Lambda\alpha \mathrm{Vb} \mathrm{H}[\bar{\alpha}(b)]}{\mathrm{H}(s)\mathrm{v} \longrightarrow \mathrm{H}(s)} \\ \Lambda b \mathrm{Q}(s\hat{\ }b) \rightarrow \mathrm{Q}(s) \\ \underline{\mathrm{H}(s) \rightarrow \mathrm{Q}(s)} \\ \mathrm{Q}(<>)$$

Remark. Ignoring the problems besetting the continuum, the monistic universes contemplated so far may be projected as a series

(H) |, ||, |||, ..., |(h), |(h'), ...

of "strokes" where h' = h + 1, i.e., the (monistic) *successor function* governing the extension of H. In É, each sequence of monistic universes has a *parametrical extension*

(H_e) |, ||, |||, ..., |(h), |(h'), ..., |(e).

where "e" is acting as the *finitization parameter*. Thus, for each sequence s of monistic universes in É there is a parameter e_s by which the projection of an infinite sequence on H is finitized by switching the projection to H_e. Consequently, I may use the notations

$$H_{e_\Lambda}, H_{e_\square}, H_{e_\blacksquare}, \ldots,$$

to denote finite images of infinite sequences of monistic universes (in É).

Furthermore, in order to get a notation for "0" some indication of the *empty stroke*, shall be used. E.g. $\emptyset$ may be used to identify such indications. Since $\mathrm{lh}(\emptyset) = 0$ i.e. the *length* of the empty stroke is zero, this additional notational convention does not increase the length of the series H or H_e (under the (recursive) convention by means of which $\mathrm{lh}(1) = 1$, and, in general, by which $\mathrm{lh}(|^k) = k$, $k = 0, 1, 2, 3, \ldots$.).

Notice that any monistic ontology based on the empty stroke has "length" zero. *Viz.* The (recursive) length of any sequence

$$E^\emptyset \quad \emptyset\emptyset\emptyset, \ldots\ldots, \emptyset\emptyset, \ldots\ldots$$

is bounded above by an infinitesimal constant $\delta > 0$, since

$$\mathrm{lh}(E^\emptyset) = 0.$$

[Also, observe that the sequence $E^\emptyset$ may be of order type η (the "Hilbert Hotel"-effect), and, if read as a single (continuous) word it exemplifies an occurrence of a *dense empty word*, i.e. a word order-isomorphic with the set Q of rational numbers, i.e. the set of all ordered pairs $<m,n>$ of integers m,n).]

The infinitesimal interval $<0, \delta_\emptyset>$ of arbitrary sequences of empty strokes is an example of an *abstract object generator* whose image is obtained as a *locally infinite space* in the interior of the environment É. I use the notation

$$E^{<0, \delta_\emptyset>}$$

to indicate the continuous image of the infinitesimal interval $<0, \delta_\emptyset>$ in the interior of É.

Remark that "$\emptyset$" is here also used as a sign for the *"empty printed event."* A blank sheet of paper may — or, may not — be considered as having printed the symbol for the empty printed event on both of its surfaces or only on one of them. The example generalizes to considerations of *stacks* of blank sheets of paper that easily yield intractable problems of identifications of occurrences of (perhaps totally undetermined) empty printed events (even apart from $\emptyset$). In the limit, the by now familiar fixed point recursion (cf. p.316)

$$\vdash^0 [\,] := [\,]$$
$$\vdash^{h+1} [\,] := \vdash(\vdash^h [\,])$$
$$\vdash^H [\,] := -[\,]$$

emerges, where $[\,]$ is chosen as symbol for the *Empty Type*.

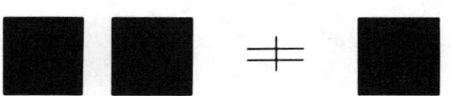

B/ W Equation

B/W Equation

B/W Equation

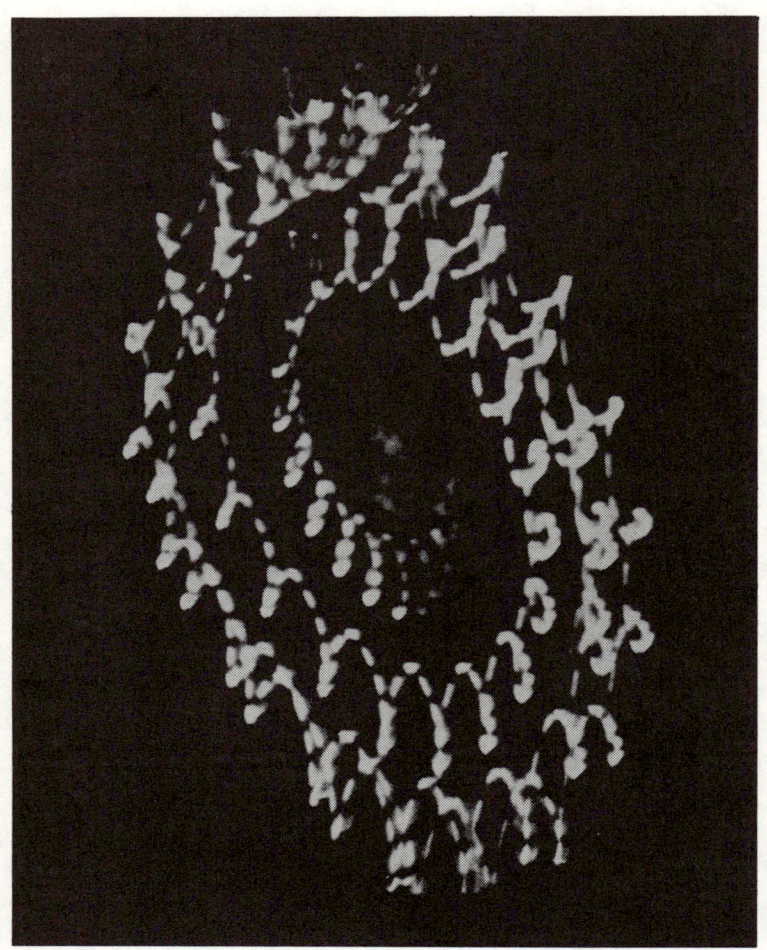

Unidentified Universe (Partial view)

Idios Kosmos (II)

II (Logo-) Graphic Forms Of The Continuum

Black Body Radiation. Planck Oscillators. (Approximations of Absolute Darkness.)

Black Topologies .... ■ ■ ....

The Chaotic Black Topology .... ■ ....

A Continuum of Black Chaotic Topologies
■ ■ ■ ■ ■ ..... ■ ■ ..... ■ ....

IIa. Indiscernables/Double-negation Topoi
IIa. Indiscernables/Double-negation Spaces.

Infinitesimal Distances/Infinitesimal Spaces.
    Infinitesimal elements are obtained by *partitions of unity,* and, as sets, contain 0 as their smallest member or minimal element. Thus, if δ is an infinitesimal, then

$$\delta \to 0$$

and

$$\delta \neq 0$$

are true relations (here "$\neq$" denotes the intuitionistic *apartness relation.*) More specifically, given a topos $\mathcal{E}$ and an object X in $\mathcal{E}$, the basic understanding when considering the topological structure determined on X by the logic of $\mathcal{E}$ is that given any two points x, y, in X, to assert the negation $\neg(x = y)$ of the equality $x = y$, means that any two points p,q are *well separated* (or at a measurable distance $\neq 0$).
    Thus, given any point p$\in X$, the points well separated from p are the points

$$x \in \neg\{p\}$$

while the points not well separated from p are the points

$$x \in \neg\neg\{p\}$$

i.e. those points x which are *infinitesimally close* (or "very close") to p.

Since, intuitionistically,

$$X = \neg\neg\{p\} \cup \neg\{p\}$$

is false in general, there must still be some space left unaccounted for in X. Remarkably, this space is a *"grey areas"* (a "no man's land") because it has *no explicit description* in the topos $\mathcal{E}$. (Remark that $\neg\neg\neg\{p\} = \neg\{p\}$ by Brouwer's argument.)

Implicitly described, the situation is as follows. Define a *neighborhood* $\mathbf{H}_p$ around a point p as a subobject $\mathbf{H} \subseteq X$ which contains $\neg\neg\{p\}$ and also this "no man's land." This is expressed by the identity

$$\neg\{p\} \cup \mathbf{H} = X$$

which says that every point in the space X is either well separated from p—or, it is in the neighborhood $\mathbf{H}_p$ of p.

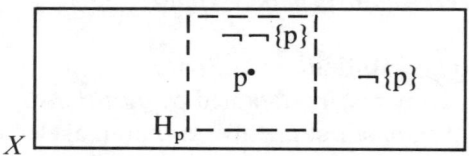

Technical Remark. By substituting for the geometrical concept of a "measurable distance" the logical concept of "provability" ("logical distance"), it is easy to obtain "topological" variants of Gödel's incompleteness proofs. *Id est,* steps of proofs are substituted for units of distance, where "proofs" of any (*independent*) axiom take exactly zero steps (i.e. are completely trivial proofs).

IIb. ■ —*Space*

I. C-*Space.* 2-*Space. Ultra-monism.*

Ia. The chaotic topology on a ■-space is given by the single non-unique arrow

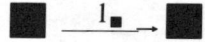

obtained by substituting every occurrence of q with p throughout all inequalities

$$p \neq q$$

The Yellow Book

including the case where $q = \neg\neg p$ i.e. where (previously)

$$\neg\neg p \neq p$$

while changing the apartness relation to a relation of identity. Set theoretically, ■ is treated as {p} i.e. as a *singleton* set. In so far as p is not unique but takes on a range of ambiguous values $p^* = p', p'', \ldots$ the singleton set {p*} is easily interpreted as a *continuously variable* (singleton) *set*.

Ib. From the chaotic topology one derives continuously variable *sets of indiscernibles* indexed by various wandering sets of distinct degrss of chaos or indeterminacy. Thus, the sequence

$$\Delta_■ \quad ■ \ldots\ldots\ldots ■\,■ \ldots\ldots ■\,■\,■\,■$$

(with (inverse) limit point □) is logographically increasing in *ultra-marine blue* color elements while continuously decreasing in *black* or *ultra-grey* color elements (until, in the limit *(via* a passage through a "grey area") *only* the boundary □ is carrying a color space (distinct from the empty color space (white elements)).

Ic. The chaotic set **C** is also interpreted as a set {0,1} or **2**-*space*, by taking 0 as denoting an empty subset of **C** while 1 denotes the entire space. I.e. the arrow

$$1 \to \mathbf{C}$$

determines the interpretation uniquely (up to isomorphism).

Id. By duality, by regarding all points $\neg\{p\}$ and $\neg\neg\{p\}$ as distinct from {p}, the ■-*space* is transformed to a **C**-*space* of the (intuitionistic) continuum requiring a *continuum of tactics of attention* to the points q distinct (at any non-zero distance) from p. [In a construction of a pair $\langle ■, ■ \rangle$ of Black Spaces of the continuum it is often convenient to regard its first projection as a constant zero-function corresponding to the empty subset of the uncountable set of points along the intuitionistic unit continuum 1_C, while its second projection takes 1 as its image — the latter, of course, denoting the *unit interval* of the continuum **C**.]

Ie. The *Ultra-black Construction*

$$E^{\langle ■,■ \rangle} \to \acute{E}$$

is a realization of the (intuitionistic) *universal spread of elements of (a species of) ultra-black ashes* as a (topological) *covering* of a pair of universal (white) squares $<\square, \square>$. The latter are represented as highly polished aluminum mirrors measuring exactly 1 m² each—a measure which except for a *meager* subset excludes the *boundaries* of the mirror surfaces which continue the covering of the interior and span a surface measuring exactly four times $10 \times .333....$ m² and identically covered by the same ultra-black particles—thereby reflecting the all-embracing effect of *The Fire in The Mirror*, an alternative title for *The Painting of a Disaster (Multiplied)*.

The latter, beginning a transfinite sequence

$$É^{\blacksquare}{}_\alpha \quad \blacksquare\ \blacksquare\ \ldots\ ,$$

is constructed from the *initial catastrophe*

$$É^{\blacksquare}\ \ldots\ \blacksquare\ \ldots\ (Black\ M\bar{A})$$

which is experienced *twice over* ($=É^{\blacksquare_2}$) under the guidance of noemas created by iterated passages through the interiors of *Black MĀ* during which a continuum of layers of darkness is traversed. *Lumen increatum. Cognitio vespertina.*

Remark. When E is realized within the context of MĀ-Theater, $M\bar{A}_{\blacksquare_{\alpha=1}}$ is also referred to as the *Initial Pine Sacrifice*. (The ash paint is obtained and applied by mixing pine-sticks with fire.)

Ie! Remark that, in contradistinction to the sequence $\Delta_\blacksquare$ of gradually varying indiscernibles ■, the *vertical topology* generated by the (ordinal) *height* of the topos is determined by the stochastic isomorphisms of an initial chaotic topology $\blacksquare_0$ (chosen by the Creative Subject.) For the purpose of visualization, the finite models of an N-stack are sometimes used as an aid for intuition.

(Ie″) In intuitionistic cosmology*, a solution to the *Night Sky Paradox* depends on both set theoretical and topological properties of the continuum C. The darkness of the night sky is indicative of the *absence* of additive groups of motion (i.e. absence of closure under a + b for arbitrary travelling elements a, b, of the light of the stars) while, simultaneously exhibiting a *logarithmic gradient of growth*—allowing for models of multiple-successor arithmetics (representing degrees of "anti-darkness".)

---

*Intuitionistic cosmology is derived from Brouwer's concept of geometry by which any cosmic (sub-)space can be obtained as the image of a *continuous group of motions* (such as Lie groups).

The Ultra-Black Paintings
(Moderna Museet, 1976)

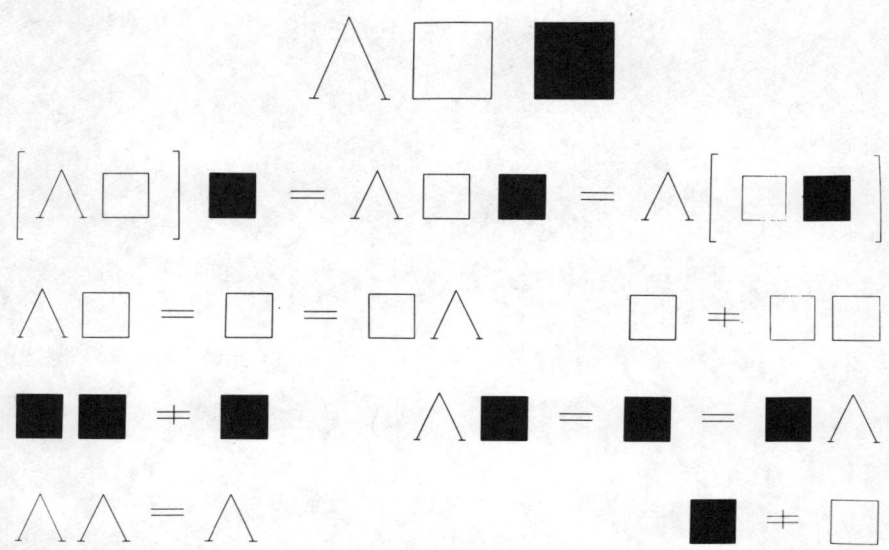

B/W Algebra

Idios Kosmos (III)
*Post-socratic Ontologies. Non-monistic Universes.
Binary Ontologies. Ternary Ontologies, ... , κ-ary Ontologies.*

(A) Dualistic Universes.

I. Binary Universes. [2-Ontologies. 2-Categories]

   0-1 - Universes: propositions as finite binary sequences

        2-valued Measures (Complete Boolean Algebras &c.)

        (Characteristic Function Universes)

        $\{\vee, \wedge\}$ — Off-On-Universes

        □ ■ -Universes

        propositions as infinite binary strings

Binary Categories: $<\rightrightarrows, \uparrow\uparrow>$

                    $<\rightrightarrows, \downarrow\downarrow>$

                    .
                    .
                    .

                    &c.

Σ contemplates an endless formation of sets

        □, Λ, ■, ΛΛ, Λ■■, ...,

coded as a binary sequence

        □, ■, ■□, ■■, ...,

with arrows

            □ → 0
            ■ → 1
            Λ → ∅

    This contemplation of ternary and binary universes arises out of a still more fundamental contemplation of a κ-dimensional intuitionistic continuum the initial spread of which may be drawn as

$$a_1 \ a_{\kappa^+} \ a_{2\kappa^+} \ \ \ldots\ldots$$
$$a_2 \ a_{\kappa^{++}} a_{2\kappa^{++}} \ \ \ldots\ldots$$

$$\ldots\ldots\ldots\ldots\ldots\ldots\ldots\ldots$$

$$a_\kappa \ a_{2\kappa} \ a_{3\kappa} \ \ \ldots\ldots$$

Using modular arithmetic the following diagram pictures a binary code for the endless progressions of *finite sets*:

	0	1	2	3	4	5	6	7	8	9	10	11	12
0	0	1	0	1	0	1	0	1	0	1	0	1	0
1	0	0	1	1	0	0	1	1	0	0	1	1	0
2	0	0	0	0	1	1	1	1	0	0	0	0	1
3	0	0	0	0	0	0	0	0	1	1	1	1	1
4	0	0	0	0	0	0	0	0	0	0	0	0	0
.	.												.
.	.												.
.	.												.
.	.												.
k	0	0	0	0	0	0	0	0	0	0	0	0	0
.	.												.
.	.												.
.	.												.

The Yellow Book

Not all sets are well-founded—in particular those preceeding Σ's initial contemplations of the universal spread of binary words in lexicographic order. Beginning with the elementary fixed-point equations

$$x = \{x\}$$
$$y = \{y\}$$
$$z = \{z\}$$

and the usual self-referential formulae, Σ is equipped to contemplate arbitrary *non-well-founded* sets such as

$$\{\ldots\{\{\ldots\{\{\{\ldots\}\}\}\ldots\}\}\ldots\},$$

again, with specific attention to the interpretation of the occurrences of lazy dots.

I notice in passing their relations of duality with the finite but indeterminate *Zermelo ordinals* **z** (z_i)

$$\{\ldots\{\{\ldots\{\{\{\Lambda\}\}\}\ldots\}\}\ldots\},$$

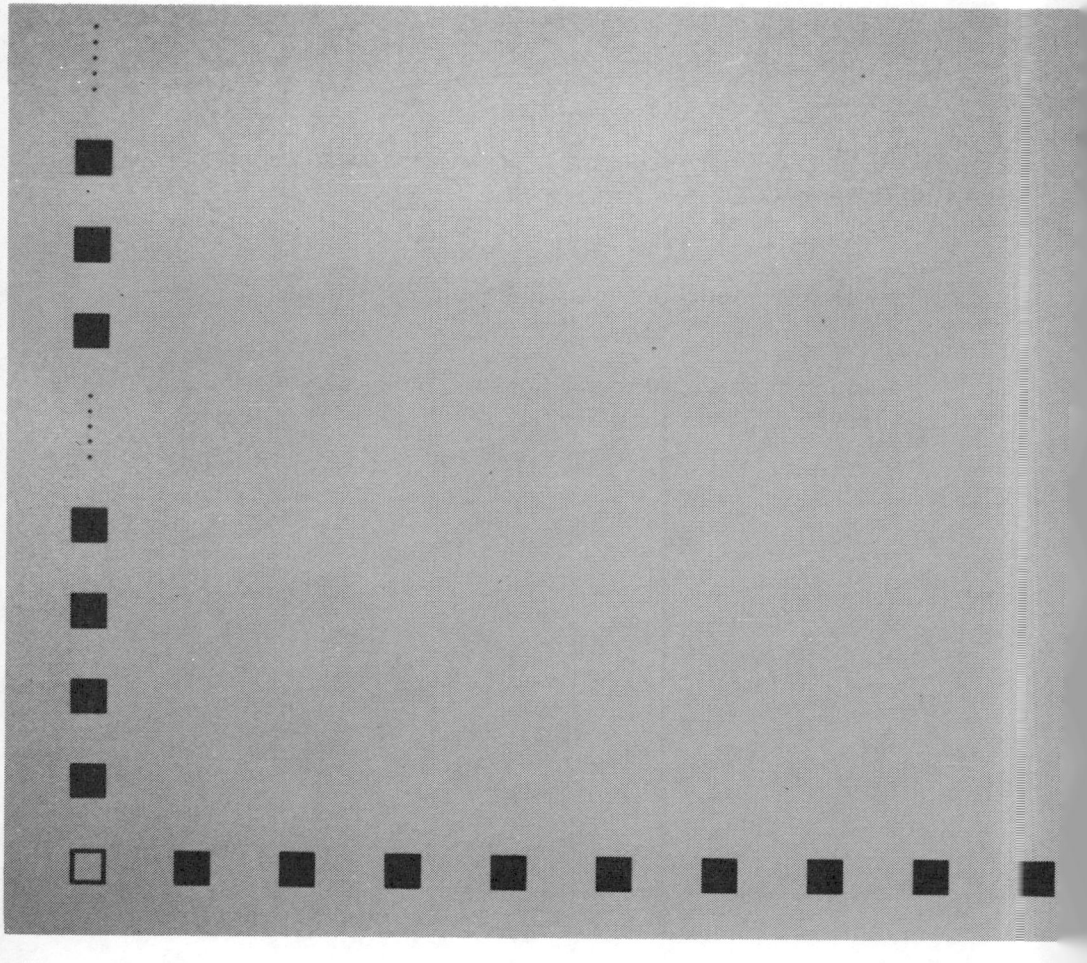

Large Infinitary Process 4 IX — 17 X 76
(Moderna Museet, Stockholm)

II. The Model Theory of Algebraic Aesthetics.
*From univalent and ambivalent truth-values to binary, ternary, ...
and κ-ary truth-values.*

~~The Lattice of Continuous Truth~~.

<u>*TACTICAL CONFIGURATION*</u>

$1 \longrightarrow 0$ (INITIAL MAP)

$0 \longrightarrow \Lambda$ (FIRST DISCRIMINATION)

$\Lambda \longrightarrow \blacksquare$ (CATASTROPHE MAP)

$\blacksquare \longrightarrow \square$ (RECONCILIATION)

$\square \longrightarrow 1$ (RETURN TO THE EVERYDAY MALICE)

.4.2: *By a "colorless" or "grey" WORLD ALGEBRA* V *we mean the notion*

$$V \approx \Lambda \square \blacksquare$$

of dialectics, where

$$\square \approx \{\Lambda\} ,$$

$$\blacksquare \approx \{\Lambda\{\Lambda\}\} ,$$

$$\Lambda \approx \overline{V} \quad and$$

$$\overline{\Lambda} \approx V$$

.4.21: EXAMPLE. $\mathbf{V} \approx \{ \Lambda \{ \Lambda \{ \Lambda \} \} \}$
together with its equational closure and tactical configuration is a WORLD ALGEBRA $\mathbf{V}$

.4.3: By a WORLD ALGEBRA $\mathbf{V}$ *we mean a WORLD ALGEBRA* $\mathbf{V}$ *containing some of the following equations*

$$\zeta : \mathbf{V} = \Lambda$$

$$\eta : \mathbf{V} = \square$$

$$\theta : \mathbf{V} = \blacksquare$$

Ad ζ. *Equation* ζ *yields* $\forall y\, A(y)$ *always <u>fulfillable</u> while* $\exists y\, A(y)$ *is always <u>unfulfillable</u>.*

Ad η. *Equation* η *yields* $\forall y\, A(y)$ *always <u>fulfillable</u> if, and only if,* $\exists y\, A(y)$ *is always <u>fulfillable</u> and conversly with unfulfillable in place of fulfillable.*

Ad θ. *Equation* θ *yields* $\forall y\, A(y)$ *<u>fulfillable</u> if, and only if,*
$\exists y\, \exists x\, [\, A(y)\, \&\, A(x)\, \&\, \neg\, y \cong x\,]$ *and* $\exists y\, A(y)$ *is always <u>unfulfillable</u> if, and only if,* $\forall y\, [\, \neg\, A(y)\, \&\, y \cong y\,]$

.4.4: <u>Definition</u>. The WORLD PROBLEM for $\mathbf{V}$ *is the class of equations each member of which has* $\square$ *as a solution.*
<u>Definition</u>. *The WORD PROBLEM for* $\mathbf{V}$ *is the class of (mental) activities* $\mathscr{AE}$ *which (in* $\square$ *) proceed over a span of indications of* Λ.

Note that in $\square$ $\Lambda\Lambda \sim \Lambda$, *id est iteration (continuation) is <u>idempotent</u>. Of course, in terms of the mundane experience of the world, WORLD and WORD PROBLEMS for* $\mathbf{V}$ *are examples of what appropriately can be called IMAGINARY PROBLEMS.*

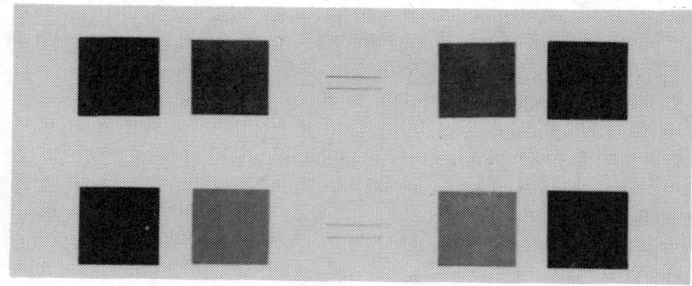

Color Equations

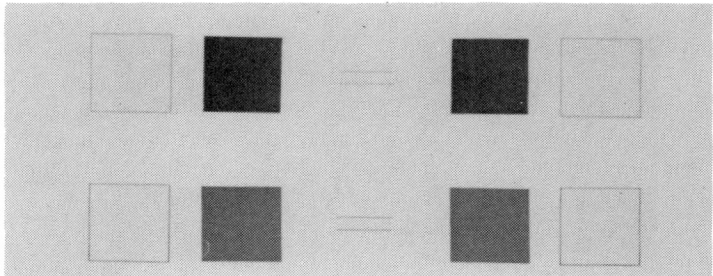

Color Equations

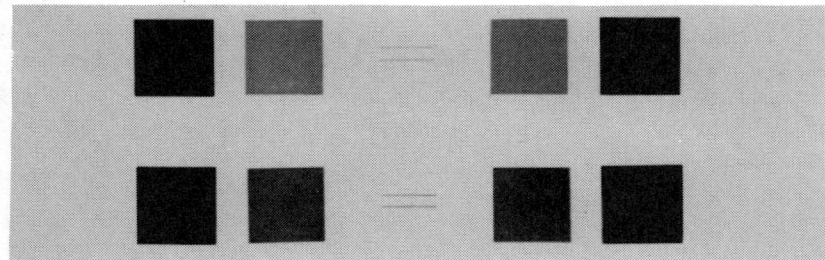

Color Equations

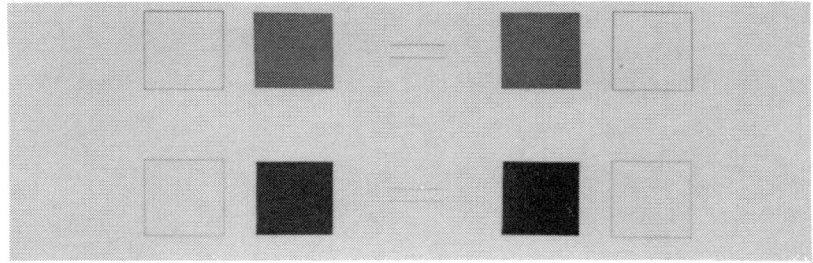

Color Equations

S E M I O T I C S - I

The idea of an <u>abstract language</u> originated with the discovery of some simple <u>algebraic properties</u> pertaining to the combinatorial or syntactical rules governing the productions of strings of symbols. One of the most elementary algebraic structures involving these properties is the <u>semi group with</u> 1, generally designated $a = \langle \mathcal{A}, \circ \rangle$, where the first coordinate designates the <u>domain</u> of the semi group (also called the carrier of the semi group) and the second coordinate designates an <u>associative operation</u> (concatenation) closed with respect to the domain of the semi group.

Operations performed on a carrier $\mathcal{A}$ are termed <u>productions</u> of the structure a and they correspond to sentence forms and other grammatically significant units on the syntactic level.

Chains of productions over $\mathcal{A}$ are given structural descriptions in terms of <u>trees</u>, $\mathcal{T}_{\mathcal{A}}$, where the "root" designates the initial production (or "start symbol") and the branchings designate the successive applications of operations on previously obtained "<u>words</u>". By the closure of $\mathcal{A}$ we mean all trees $\mathcal{T}_{\mathcal{A}}$ such that $\mathcal{T}_{\mathcal{A}}$ obtains in a. This closure corresponds to the <u>language</u> generated by a, designated $\mathcal{L}(\mathcal{A})$.

By generalizing the concept of a semi group with 1 we may obtain a global presentation of $\mathcal{L}(\mathcal{A})$. This generalization brings us to the concepts of <u>category theory</u>, one of the highlights of exact thinking after the creation of Cantor's "Paradise". By a <u>category</u> $\mathcal{C}$ we shall understand a collection of <u>objects</u>, corresponding to the words on $\mathcal{A}$ above, together with a collection of <u>morphisms</u>, corresponding to the productions generating $\mathcal{L}(\mathcal{A})$.

Objects will be denoted $A, B, Γ, Д, \ldots$ and morphisms $Я, Ю, Э, \ldots$. For each pair of objects A, B, there is a set $C(A, B)$ of morphisms $Я$ carrying A to B, and, in addition, for each A in C, an <u>identity morphism</u> $\epsilon_A(A, A)$, also written $\mathcal{I}_A$.

Now, if there are three morphisms $Я, Ю, Э$ such that
$$Я : A \Rightarrow B, \quad Ю : B \Rightarrow Γ, \quad Э : Γ \Rightarrow Д$$
then the <u>composition</u> of them satisfies

$$(ЯЮ)Э = Я(ЮЭ)$$

Also, if $Я : A \Rightarrow B$, then $\mathcal{I}_A Я = Я = Я \mathcal{I}_B$

To complete our definition we remark that a Category C is always <u>closed</u> under arbitrary <u>compositions</u> of morphisms, i.e. if

$$Я \in C(A, B) \text{ and } Ю \in C(B, Γ), \text{ then always}$$

$$ЯЮ \in C(A, Γ)$$

The <u>syntax</u> of $\mathcal{L}(A)$ can now be specified as a category $\mathcal{F}$ where the objects are strings of letters drawn from some fixed alphabet and the morphisms are derivations (trees) of one string from another.

By a derivation $\mathcal{D}$ we shall now understand the following ordered triple

$$\mathcal{D} = \Big\langle (A_0, \ldots, A_n), (Я_0, \ldots, Я_{n-1}),$$
$$(\lambda_0 - \zeta_0, \ldots, \lambda_{n-1} - \zeta_{n-1}) \Big\rangle$$

WHERE

$(A_0, \ldots, A_n)$ DESIGNATES THE <u>WORD COORDINATE</u>,

$(Я_0, \ldots, Я_{n-1})$ IS THE <u>DERIVATION COORDINATE</u>, AND

$(\lambda_0 - \zeta_0, \ldots, \lambda_{n-1} - \zeta_{n-1})$ IS THE <u>NEIGHBOURHOOD COORDINATE</u>,

THE LATTER GIVEN A <u>TOPOLOGICAL</u> INTERPRETATION (SEE BELOW).

THE <u>LENGTH ZERO</u> DERIVATION $\langle (A), (\ ), (\ ) \rangle$ IS REGARDED AS THE A-<u>IDENTITY DERIVATION</u>, WHILE THE <u>LENGTH ONE</u> DERIVATION $\langle (A, B), (Я : A \Rightarrow B), (\lambda - \zeta) \rangle$ WILL BE "ABBREVIATED" $Я : A \Rightarrow B$ AND PURPOSELY CONFUSED WITH THE "REAL" PRODUCTION $A \Rightarrow B$ IN $\mathcal{L}(A)$. CLEARLY $\mathcal{D}$ MAY BE TURNED INTO AN INDEPENDENT CATEGORY, ID EST INDEPENDENT OF $\mathcal{L}(A)$.

THE IMPORTANCE OF THE ABOVE CONCEPTS COMES FROM THE FACT THAT MORPHISMS ARE EXAMPLES OF A GENERAL CLASS OF <u>ARROWS</u>. ANOTHER EXAMPLE IS THE CLASS OF <u>FUNCTORS</u> THAT EXISTS BETWEEN CATEGORIES THEMSELVES.

A FUNCTOR Υ FROM A CATEGORY C_0 TO A CATEGORY C_1 IS SIMPLY TWO CLASSES OF ARROWS, ONE SENDING OBJECTS IN C_0 TO OBJECTS IN C_1 AND THE OTHER SENDING MORPHISMS IN C_0 TO MORPHISMS IN C_1. FORMALLY, THE FOLLOWING SITUATION OBTAINS: IF Υ IS A FUNCTOR BETWEEN C_0 AND C_1, ID EST

$$\Upsilon : C_0 \longrightarrow C_1$$

, THEN FOR EVERY $A \in C_0$, Υ ASSIGNS AN OBJECT $\Upsilon(A)$ IN C_1 AND FOR EACH $Я \in C_0$ Υ ASSIGNS THE MORPHISM $\Upsilon(Я)$ IN C_1.

Now, by the SEMANTICS of $\mathcal{L}(\mathcal{A})$ we mean a CO-FUNCTOR

$$\mathcal{J} : \mathcal{F} \rightrightarrows \mathcal{U}.$$

Where $\mathcal{U}$, of course, denotes the SEMANTIC CATEGORY associated with $\mathcal{L}(\mathcal{A})$ when generated by the Syntax Category $\mathcal{F}$.

More specifically, the Co-Functor $\mathcal{J}$ specifies an INTERPRETATION of $\mathcal{F}$ by taking Objects to CARTESIAN PRODUCTS in $\mathcal{U}$ and derivations to functions in $\mathcal{U}$. In other words, the Semantic Category $\mathcal{U}$ consists of a Category of SETS and FUNCTIONS and the image of the interpretation $\mathcal{J}$ is called the SEMANTICS of the interpretation. For example, the interpretation of an Object $\mathbf{A}_\mathcal{F}$ of the Syntax Category consists of those Functions Ш that are Contravariant in $\mathcal{U}$. (They correspond to retracts in a Topos).

If Я is a Morphism in $\mathcal{U}$ (i.e. Я is a Function), then the neighbourhood of Я is defined as those Morphisms Я₁, ..., Яₙ that act as Identities on the extended neighbourhood domain of Я. That is, Я₁, ..., Яₙ are neighbourhoods of Я if the following diagrams commute:

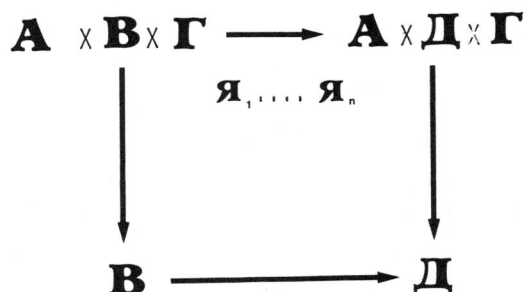

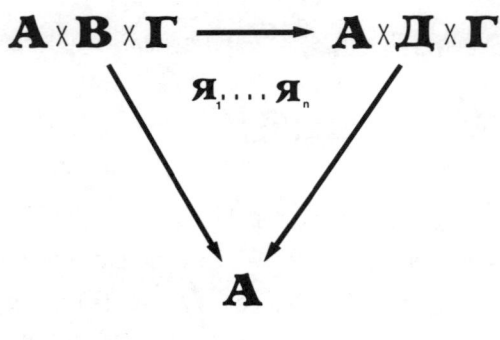

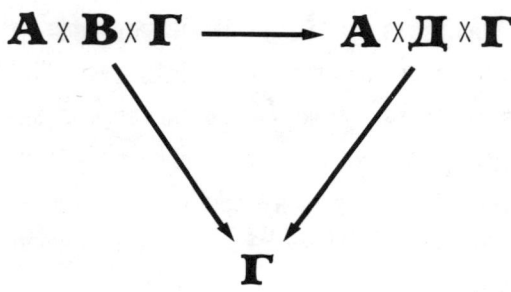

where **Я** is the morphism **Я**:**B** ⇒ **Д** and X indicates Cartesian Product.

Further examples of general classes of Arrows to be met with in Category Theory - besides Morphisms and Functors - are <u>MONO-MORPHISMS</u> and <u>EPI-MORPHISMS</u>, as in the following

<u>DEFINITION</u>. **Щ** is a <u>MONO-MORPHISM</u> in a Category **C**,
if for all **Ю,Э** ∈ **C**

Ю:**Г** ⇒ **A** , **Э**:**Г** ⇒ **A**

implies **Ю** = **Э**   if **ЮЩ** = **ЭЩ**

DEFINITION. Ш IS AN EPI-MORPHISM IN A CATEGORY **C**,
IF FOR ALL **Ю, Э** ∈ **C**

$$Ю : B \Rightarrow Г, Э : B \Rightarrow Г$$

IMPLIES $Ю = Э$ IF $ШЮ = ШЭ$

Ш AND Ш WILL BE OF IMPORTANCE FOR THE MAPPINGS OF THE INNER STRUCTURES OF TOPOSES.

TOPOSES GIVE RISE TO YET OTHER GENERAL CLASSES OF ARROWS LIKE <u>PULL-BACKS</u> AND <u>PUSH-OUTS</u>. FOR THE MOMENT WE WILL STOP THE PRESENT CLASSIFICATION OF ARROWS, ONLY MENTIONING ONE FURTHER CLASS, AND LEAVING THE OTHERS FOR ANOTHER OCCASION. THIS CLASS IS CALLED THE <u>CLUB OPERATORS</u>, Ω, WHICH ASSIGNS TO EACH OBJECT **A** A MORPHISM

$$Ч_A : \Upsilon \mathbf{A} \longrightarrow \Phi \mathbf{A}$$

WHERE Υ AND Φ ARE **2**-FUNCTORS.

FURTHER, TO EACH MORPHISM Ч OF THE UNDERLYING **2**-CATEGORY **Г** OF **Д**, A **2**-CELL MORPHISM $Ч_A$ IN THE CO-DOMAIN **2**-CATEGORY **Д** IS ASSIGNED SUCH THAT THE FOLLOWING "SQUARE" OBTAINS:

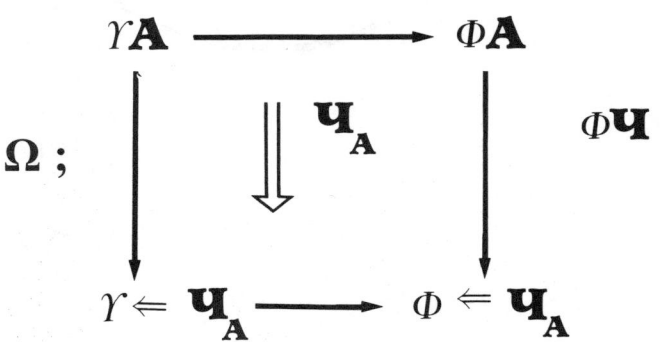

Intuitively, Ω is a <u>NATURAL TRANSFORMATION</u> between Categories. Specific Natural Transformations will be pictured:

$$\text{Ш}_\Omega : \Upsilon \rightsquigarrow X$$

The Club Operator plays a prominent role for the <u>PASSAGE</u> between Categories, as when we wish to go from one Semantic Category $\mathcal{U}_1$ to another Semantic Category $\mathcal{U}_2$, with common underlying Syntactic Category $\mathcal{F}_0$. The DEONTIC AXIOMS determining the admissible passages for some Ω are called the <u>DOCTRINE OF THE CLUB OF CATEGORIES</u> (or just <u>DOCTRINES FOR SCHOOLS</u>), where the "Club" notion now refers to the Categories in the domain and co-domain of Ω. For example, a situation that is typical for any Doctrine for Schools is the operation of <u>PASTING</u> Objects (here, categories) in a Club. So one thing permitted is to pass from the situation

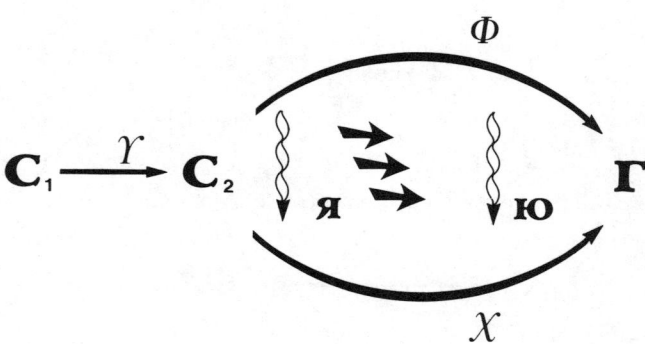

TO THE SITUATION

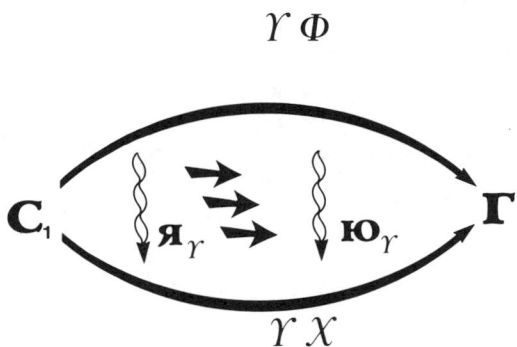

WHERE ⇉ DENOTES THE PASTING OPERATION AND Γ IS AN ELEMENT OF THE CLUB.

DIAGRAMS OF CONCEPT FORMATION PROCESSES UNDERLYING FORESTS OF CORRECT REASONINGS MAKE HEAVY USE OF ⇉ , ESPECIALLY WHEN THE CLUB HAS MANY-SORTED DOCTRINES FOR SCHOOLS.

+ + + + +

SOME ELEMENTARY CATEGORIES (ARROW AND TIME CATEGORIES).

(1) LET $\mathcal{V}$ BE A (SMALL) UNIVERSE OF <u>SETS</u>. THE <u>ARROW CATEGORY</u> $\mathcal{V}^{\Rightarrow}$ WILL CONSIST OF THOSE OBJECTS $\lambda : \mathbf{X}_0 \Rightarrow \mathbf{X}_1$ FOR WHICH $\mathbf{X}_0$ IS A DOMAIN FOR λ AND $\mathbf{X}_1$ ITS

CO-DOMAIN AND WHERE THE MORPHISMS, I.E. ARROWS

$$\Lambda : \lambda : \mathbf{X}_0 \Rightarrow \mathbf{X}_1 \Rightarrow \lambda' : \mathbf{Y}_0 \Rightarrow \mathbf{Y}_1 ,$$ ARE

THE PAIRS OF FUNCTIONS λ_0, λ_1 SUCH THAT THE FOLLOWING DIAGRAM COMMUTES, I.E. $\lambda \Lambda^0 = \Lambda' \lambda$

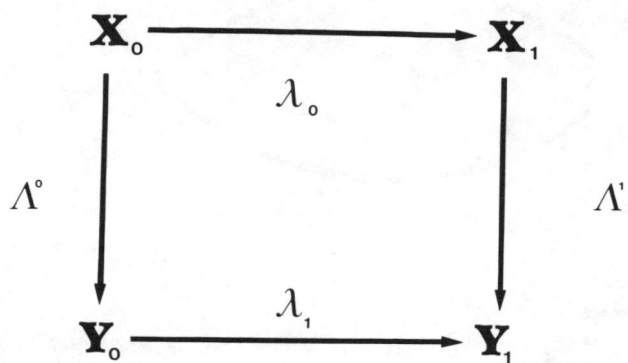

(2) IF $\mathcal{P}$ IS A (DISCRETE) <u>PROCESS</u>, THEN $\mathcal{V}^\mathcal{P}$ IS THE CATEGORY OF TIMES (RELATIVE $\mathcal{P}$). IF INSTEAD OF $\mathcal{V}$ AND/OR $\mathcal{P}$ WE PUT THE CATEGORY ON THE FORMS

$$\mathcal{V}^\mathcal{E} \quad \text{OR} \quad \mathcal{E}^\mathcal{E} ,$$

WE GET THE (UNIVERSAL) CATEGORIES OF LOCAL AND GLOBAL TIME, RESPECTIVELY.

FOR THE CATEGORY $\mathcal{V}^\mathcal{P}$ WE SHALL HAVE AS OBJECTS INFINITELY PROCEEDING SEQUENCES OR STRINGS

$$\mathbf{X}_0 \Rightarrow \mathbf{X}_1 \Rightarrow \mathbf{X}_2 \Rightarrow \cdots \Rightarrow \mathbf{X}_m \Rightarrow \cdots$$

OF ARROWS BETWEEN SETS $\mathbf{X}_j \in \mathcal{V}$ AND AS MORPHISMS SEQUENCES OF Λ-ARROWS, LIKE THE FOLLOWING:

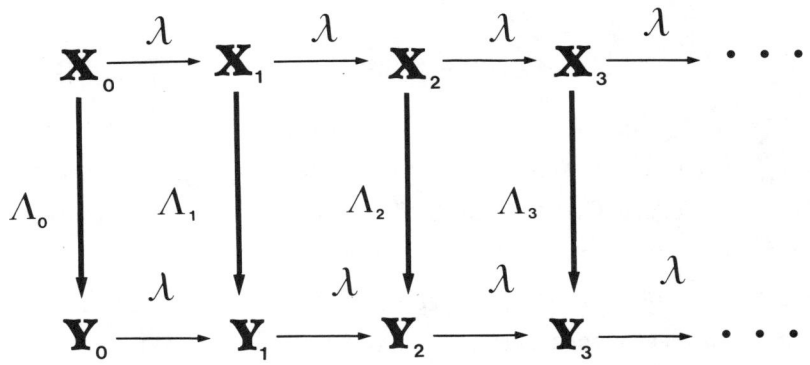

IN TERMS OF ARROWS, $\mathcal{V}^{\Rightarrow}$ AND $\mathcal{V}^{P}$ DIFFER IN THE FOLLOWING WAY

$$\mathcal{V}^{\Rightarrow} : \cdot \Rightarrow \cdot$$
$$\mathcal{V}^{P} : \cdot \Rightarrow \cdot \Rightarrow \cdot \Rightarrow \cdots$$

THE DEFINITIVE CATEGORY OF ALL CATEGORIES.

FOR TOPOSES AS WELL AS ELSEWHERE, THE DOCTRINAL CATEGORY $\mathcal{E}$ IS THE <u>FORGETFUL CO-DOMAIN CATEGORY</u> OF THE <u>UNIVERSAL FORGETFUL FUNCTOR</u> $\mathcal{E}^{\#}$

ITS OBJECTS ARE MANY-SORTED – BRACKETS, $|\ |$, AND CUPS, $\cup$ – AND THE MORPHISMS ARE OBTAINED THROUGH ITERATIONS IN THE CUMULATIVE HIERARCHY $\mathcal{V}$ OF ARROW OPERATIONS $\rightarrowtail$ FOR DENOTATIONAL CONNECTIONS BETWEEN OBJECTS IN $\mathcal{E}$.

ANY DIAGRAM FOR A CONCEPT FORMATION PROCESS (GUIDED BY CORRECT REASONINGS) HAS AN <u>ADJOINTED FORGETFUL CO-FUNCTOR</u>

MAPPING ELEMENTS IN THE PROCESS TO CARTESIAN PRODUCTS OF OBJECTS IN $\mathcal{E}$ AND MORPHISMS TO FUNCTIONS (ARROWS) IN $\mathcal{E}$. EVIDENTLY, CARTESIAN CLOSED CATEGORIES (CCC's) WILL BE THE MOST PROMINENT TOOL FOR CONSTRUCTING TOPOSES. BY THIS LAST STEP WE ATTEMPT A MOVE FROM LAWVERE'S <u>OBJECTIVE DIALECTICS</u> TO <u>NATURAL DIALECTICS</u> WHERE CLUB OPERATIONS AND OTHER NATURAL TRANSFORMATIONS (∿∿∿▶) DOMINATE OVER THE SPECIFICATIONS OF CATEGORY OBJECTS. CLEARLY, $\mathcal{E}$ CONSIDERED AS A CATEGORY WILL BE OUR ALTERNATIVE TO THE CATEGORY OF ALL CATEGORIES AS A FOUNDATION OF MATHEMATICS AND GENERALIZED CONSTRUCTIVE CONCEPTUALISM.

OF COURSE, $\mathcal{E}$ AND ITS SUB-OBJECTS WILL STILL SERVE AS OUR FAVORITE EXAMPLE OF INACCESSIBLE CARDINALS, TOPOSES, COSMOIS AND WHAT NOT THAT WILL BE ENCOUNTERED IN THE PURSUIT OF THE DELIGHTS OF EXACT THINKING.

RECALL THAT THE OBJECTS OF THE DOCTRINAL CATEGORY $\mathcal{E}$, I.E. THE DENOTATIONAL CONNECTIONS BETWEEN SIGNS OF TYPE $\left|\mathbf{a}_i\right|^k$ OR $\sqcup (\left|\mathbf{a}_{i_1}\right|^{k_1}, \ldots, \left|\mathbf{a}_{i_n}\right|^{k_m})$ CONSIST OF THE POINTS OF INDICATIONS CONNECTED WITH THE LOCAL ARROW OR PRE-MORPHISM APPEARing in some fiber for a Site belonging to the Topos $\mathcal{E}$. The Rank or Depth of any such sign is given by its associated INDEX, i.e. $i+k$ or $i+k+\frac{n\pm m}{2}$, as the case may be. In other words, the Depth of a sign $[\mathbf{a}_i]^k$ or $\sqcup([\mathbf{a}_{i_1}]^{k_1}, \ldots, [\mathbf{a}_{i_n}]^{k_m})$ equals the number of Arrows or Pre-Morphisms of the underlying Bundle or single Fiber that defines the corresponding indications.

The Depth of an indication, in turn, is generally equal to the inverse of the Rank of the Site or Stack that envelopes the indication. Analogously, the Depth of a Topos is in general equal to the inverse of the Rank of its Sheaf and so on for the remaining structures (Fiber Bundles, pb's, pc's, Sheaves, etc.).

In the other direction we may Reflect upwards over the entire universe of Sheaves. As a first stage of this Reflexion, we arrive at a PRE-COSMOI, while at a later stage the appearance of a COSMOI of collections of sets of Sheaves becomes possible.

For Cosmois, we may define LEFT and RIGHT ADJOINTS as a Universal Property of every Cosmoi. That means in particular that every Cosmoi is symmetrically SELF-REFLEXIVE with respect to ADJOINTNESS.

Adjoints associated with Sheaves and Toposes are not necessarily SYMMETRIC, but nevertheless defined in Left or Right form for every Sheaf and Topos. If both forms are permitted by the Doctrine on a Sheaf or Topos, the Universal Property is recovered. Such situations are referred to as DOCTRINAL CLUB CONSPIRATIONS - DCC's - and their Objects are the Adjointed Sheaves and Toposes.

In this way we can continue to Reflect upwards beyond the Cosmois towards more and more comprehensive Units and there seems to be no conceivable end to the iterative applications of the <u>Reflexion Principle</u>. On the other hand, due to the normality of our Sheaves, there is no corresponding infinite regress or descent downwards, since all structures below the Sheaves collapse ultimately on the void indication $\mathbf{a}_o$, i.e. the Universal Terminal Object $\underset{\sim}{\Pi}^o$. This property will henceforth be referred to as the well-foundedness of the structure in question. On the contrary, regressive Sheaves or Toposes can never be well-founded and the same holds for their (inadmissible) Doctrines.

$$\underline{\Delta(\Sigma)}$$

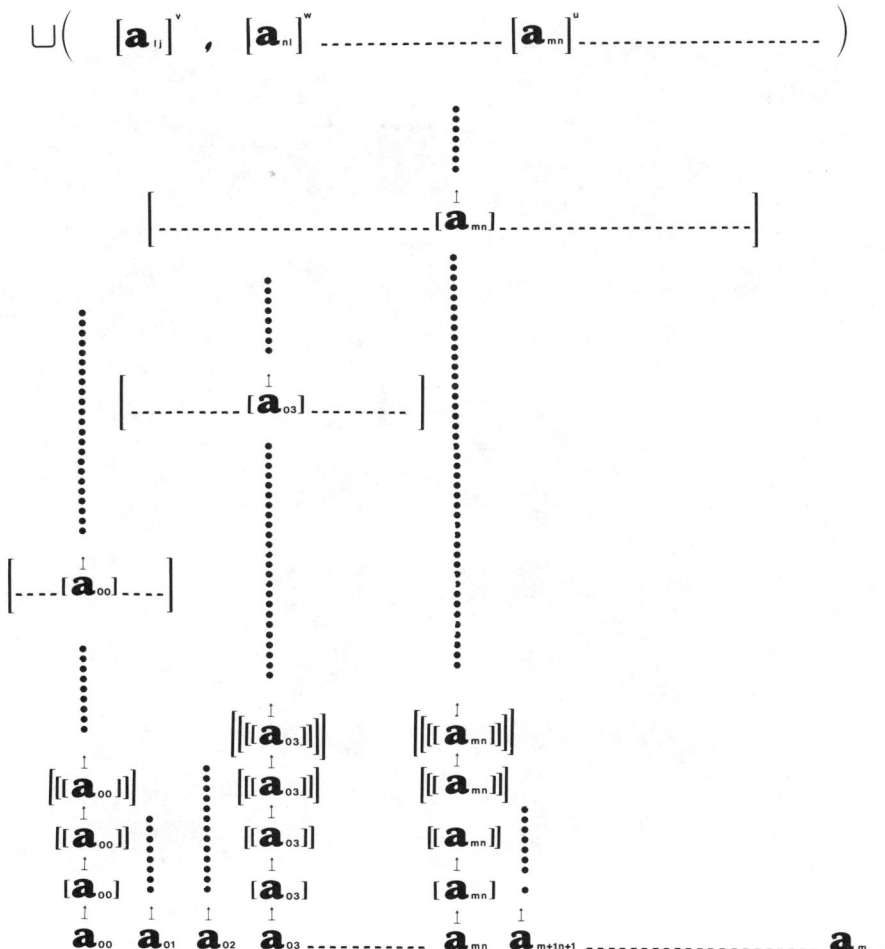

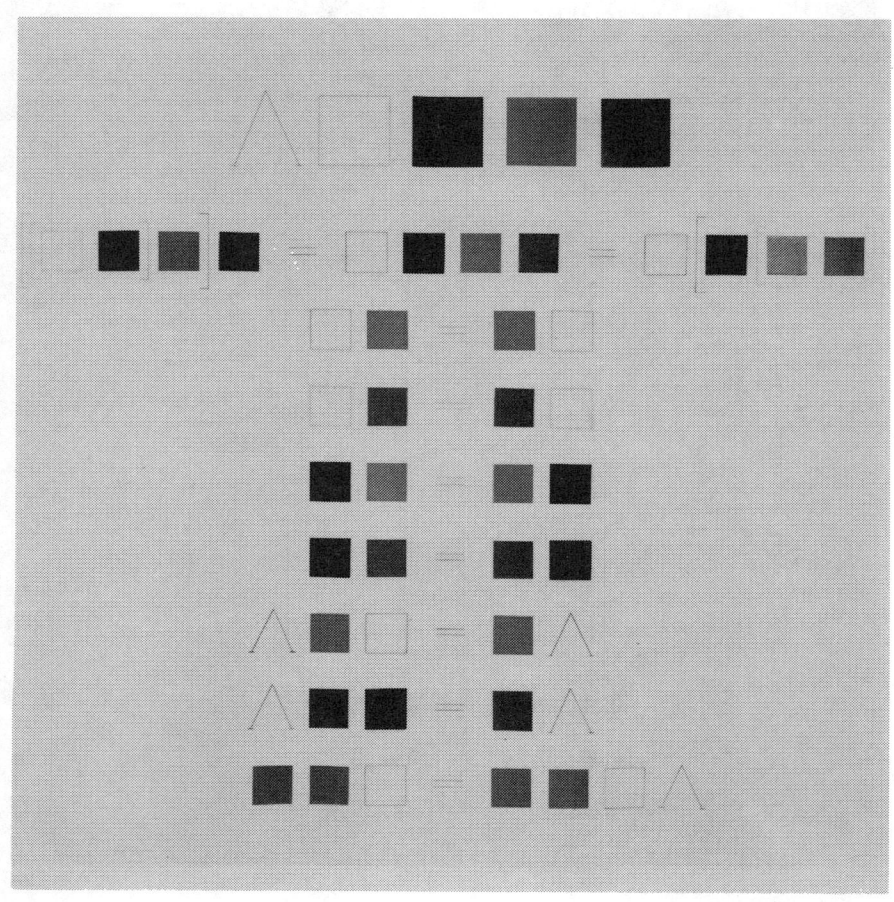

Color Algebra

Appendix I
Abandoned Lecture Notes on Intuitionism
(From the Rheinbeck Seminar, 1984)

TEXTS By Luitzen Egbertus Jan Brouwer

From L.E.J. Brouwer: *Begrundung der Mengenlehre unabhängig von logischen Saltz vom ausgeschlossenen Dritten* (Collected Works)

Der Mengenlehre liegt eine unbegrenzte Folge von Zeichen zu grunde, welche bestimmt wird durch ein erstes Zeichen und das Gesetz, das aus jedem dieser Zeichen das nachtfolgende herleitet. Unter den mannigfachen hierzu brauchbaren Gesetzen erscheint dasjenige am geeignetesten, welches die Folge ζ der Ziffernkomplexe 1,2,3,4,5,..... erzeugt.

Die Bestimmungsgesetze endlicher Zeichengruppen sowie unbegrenzter Zeichenfolgen von der Art der Folge ζ, bilden besondere Falle von Mengen, deren Elemente von der einzelnen Zeichen gebildet werden. Die Menge der Ziffernkomplexe von ζ werden wir mit A bezeichen.

Mengen und Elementen von Mengen werden *mathematische Entitäten* genannt.

Eine Species E heisst *endlich*, wenn sie mit der Menge der Ziffernkomplex eines gewissen Anfangselementes der Folge ζ gleichmächtig ist.

Eine Species U heisst *unendlich*, wenn jedes Element von A einem verschiedenen Elemente w von U, zugeordnet werden kann. Im Falle dass die Elemente w eine mit A gleichmächtige abtrennbare Teilspecies von U bilden, heissr U *reduzierbar unendlich*.

Es existiert kein Grund zu Behaupten, dass jede Menge oder Species entweder endlich oder unendlich sei. Dagegen steht fest, dass eine Species nich gleichmächtig endlich oder unendlich sein kan und zwar auf Grund des folgenden Satzes:

The Yellow Book

Haupteigenschaft der endlichen Species. Fur jede Herstellungsweise der eineindeutigen Bezeichnung zwischen einer endlichen Species E und der Menge der Ziffernkomplexe eines Anfangselementes von ζ, kurtz: fur jedes Zählungsweise von E, wird dasselbe Anfangselemente von ζ benutz.
. . . . .

Die Species C_n der Gruppen von n unbeschränkt fortgesetzten Folgen von zu ζ gehörigen Ziffernkomplexen ist eine Menge derselben Kardinalzahl wie die Menge C. Um dies einzusehen, braucht man nur dem Elemente $a_1\ a_2\ a_3\ a_4 \ldots$ von C das Element

$$a_1\ a_{n+1}\ a_{2n+1} \ldots$$

$$a_2\ a_{n+2}\ a_{2n+2} \ldots$$

. . . .

$$a_n\ a_{2n}\ a_{3n} \ldots$$

von C_n zuzuordnen. In dieser Weise bestimmt man gleichzeitig eine eineindeutige Beziehung zwischen den Punkten eines n-dimensionalen Kubus und den Punkten eines geraden Liniensegmentes (aus welcher unmittelbar eine eineindeutigee Beziehung zwischen den Punkte der geraden Linie folgt).

Dieser Beziehung ist aber *nich stetig* wenn man z.B. (bei konstanten $a_1, \ldots, a_n, a_{n+3}, a_{n+4} \ldots$.) a_{n+2} abwechselnd gleich 1 und 2 wählt und a_{n+1} unbeschränkt wachsen lässt, so bekommt man auf dem Liniesegmente eine gegen einen einzigen Punkt konvergierende Folge von Punkten, im n-dimensionalen Kubus aber *nicht* gegen einen einzigen Punkt konvergiertende Folge von Punkten.

The Two Basic Acts of Intuitionism

by L.E.J. Brouwer

From L.E.J. Brouwer: *Historical Background Principles and Methods of Intuitionism* (Collected Works)

......*The intervention of intuitionism* of which the first seems necessarily to lead to destructive and sterilizing consequences, whereas the second yields ample possibilities for recovery and new developments......

FIRST ACT OF INTUITIONISM
completely separates mathematics from mathematical language, in particular from the phenomena of language which are described by theoretical logic. It recognizes that mathematics is a languageless activity of the mind having its origin in the basic phenomenon of perception of a *move of time*, which is the falling apart of a life moment into two distinct things, one of which gives way to the other, but is retained by memory. If the two-ity thus born is divested of all quality, there remains the common substratum of all two-ities, the mental creation of the *empty two-ity*. This empty two-ity and the two unities of which it is composed, constitute the *basic mathematical systems*. And the basic operation of mathematical construction is the *mental creation of the two-ity of two mathematical systems previously acquired*, and the consideration of this two-ity as a new mathematical system.

It is introspectively realized how this basic operation, continually displaying *unaltered* retention by memory, successively generates each natural number, the infinitely proceeding sequence of natural numbers, arbitrary finite sequences and infinitely proceeding sequences of mathematical systems previously acquired, finally a continually extending stock of mathematical systems corresponding to "separable" systems of classical mathematics.

THE SECOND ACT OF INTUITIONISM

which recognizes the possibility of generating new mathematical entities:

firstly in the form of *infinitely proceeding sequences* $p_1\ p_2\ldots$, whose terms are *chosen more or less freely from mathematical entities previously acquired;* in such a way that the freedom of choice existing perhaps for the first element p_1 may be subjected to a lasting restriction at some following p_ν, and again and again to sharper lasting restrictions or even abolition at further subsequent p_ν's, while all these restriction interventions, as well as the choices of the p_ν's themselves, at any stage may be made to depend on possible future mathematical experiences of the creating subject;

secondly in the form of mathematical species, i.e. *properties supposable for mathematical entities previously acquired,* and satisfying the condition that, if they hold for a certain mathematical entity, they also hold for all mathematical entities which have been defined to be equal to it, relations of equality having to be symmetric, reflexive and transitive; mathematical entities previously acquired for which the property holds are called *elements* of the species.

With regard to this definition of species we have to remark firstly that, during the development of intuitionist mathematics, some species will have to be considered as being re-defined time and again in the same way, secondly that a species can very well be an element of another species, but never an element of itself.

Two mathematical entities are called *different* if their equality has been proved to be absurd.

Two infinitely proceeding sequences of mathematical entities a_1, $a_2,\ldots$, and $b_1, b_2,\ldots$, are called *equal* or *identical* if $a_\nu = b_\nu$ for each ν, and *distinct,* if a natural number s, can be indicated such that a_s and b_s are different.

Essentially Negative Properties
by L.E.J. Brouwer

In order to estimate the significance of affirmative or negationless mathematics, the development of which is sometimes advocated, it may be useful to publish a simple and clear example, which I gave now and then in courses and lectures since 1927, and in which a simply negative property (i.e. the absurdity of a constructive property) is realized in such a way that no hope of transforming it into a constructive property can be justified. It consists of two real numbers which are different though neither can be proved to be greater or smaller than the other, let alone that they can be proved to be apart from each other.

Let α be a mathematical assertion that *cannot be tested,* i.e. for which no method is known to prove either its absurdity or the absurdity of its absurdity.[1]

Then the creating subject can, in connection with the assertion α, create an infinitely proceeding sequence of rational numbers $a_1, a_2, a_3, \ldots$ according to the following direction: As long as, in the course of choosing the a_n, the creating subject has experienced neither the truth, nor the absurdity of α, every a_n is chosen equal to 0. However, as soon as between the choice of a_{r-1} and that of a_r the creating subject has obtained a proof of the truth of α, a_r as well as $a_{r+\nu}$ for every natural number ν is chosen equal to 2^{-r}. And as soon as between the choice of a_{s-1} and that of a_s the creating subject has experienced the absurdity of α, a_s as well as $a_{s+\nu}$ for every natural number ν is chosen equal to -2^{-s}.

This infinitely proceeding sequence $a_1, a_2, a_3, \ldots$ is positively convergent, so it defines a real number ρ.

If for this real number ρ the relation $\rho > 0$ were to hold, then $\rho < 0$ would be impossible, so it would be certain that α could never be proved to be absurd, so the absurdity of the absurdity of α would be known, so α would be tested, which it is not. *Thus the relation $\rho > 0$ does not hold.*

---

[1] For instance the assertion that there exists a quadruple of natural numbers 2,a,b and c, for which the relation $a + b = c$ holds, or that in the decimal expansion of Π there occur ten successive digits forming a sequence 0123456789.

Further, if for the real number ρ the relation $\rho<0$ were to hold, then $\rho>0$ would be impossible, so it would be certain that α could never be proved to be true, so the absurdity of α would be known, so again α would be tested, which it is not. *Thus neither does the relation $\rho<0$ hold.*

Finally let us suppose that the relation $\rho=0$ holds. In this case neither $\rho>0$ nor $\rho<0$ could ever be proved, so neither the absurdity nor the truth of α could ever be proved, so the absurdity as well as the absurdity of the absurdity of α would be known. This is a contradiction, *so the relation $\rho=0$ is absurd,* in other words *the real numbers ρ and 0 are different.*

Consequently for the real numbers ρ and 0 the simply negative property $\rho\neq 0$ holds, whilst neither of the properties $\rho\gtrless 0$ or $\rho\lessgtr 0$ is present, let alone one of the constructive properties $\rho\gtrless 0$ or $\rho\lessgtr 0$. Thus for real numbers the relation $\neq$ is an essentially negative relation.

Analogously, if at the end of the third section above, -2^{-s} is replaced by 2^{-s}, then for the real numbers ρ and 0 the simply negative property $\rho>0$ holds, while the constructive property $\rho\gtrless 0$ does not hold. Thus the relation $>$ of virtual order is also an essentially negative relation.

Lecture Notes on Intuitionism (I)

Cognitio enim contingit secundum quod cognitum est cognoscente. Cognitum autem est in cognoscente secundum modum cognoscentis. Unde cuiuslibet cognoscentis cognito est secundum modum suae naturae.

Th(Σ) is the intuitionistic theory of *initiable* mathematical realities. This theory contains many possibilities but they are all characterized by being *bounded* by what is *presently* given.*

Since what is present to the creative subject, Σ, *varies* with the stages of her illumination, it is by studying the very variations of these stages that we can grasp what kind of possibilities Σ is *limited* by.

For limited she is at any stage of illumination, be it after $\phi(n)$ lives or not. But even so, we can agree that Σ is *never less* illuminated after the nth illumination than she is after her first initiation with the practice of a single successor function λ_{00}, that is, after her very first illumination (by an unending series of natural numbers).

The theory of the creative subject, Th(Σ), derives its entire rationale from the assumption that iterations of illuminations initiate Σ to grasp still higher stages of illumination.

. . . . .

We begin by considering an indefinitely proceeding sequence

$$\underline{\mathcal{U}} : \mathcal{U}_a, \mathcal{U}_b, \ldots\ldots, \mathcal{U}_k, \mathcal{U}_{k+1}, \ldots\ldots$$

of *universes*.

Inside a universe $\mathcal{U}$ there emerges two kinds of events, viz those which are judged to hold for the *fleeing properties* of $\mathcal{U}$ and those stable or non-fleeing properties which hold for the already arrived events together with their lawful (functional) continuations within the limits of $\mathcal{U}$.

---

*The immanently derivative character of symbolic assertions stems from the impossibility of a presuppositionless interpretation of occurrences of symbolic events.

Hence, all symbolic actions are lost in their own opacity which individual assertions consecrate as a kind of ideological subjective camouflage in lieu of the apodictic evidence for the aberrance of Being.

As a result, isolation becomes a prerequisite as a means of communication as does the need for undoing what has been instrumental for the reification of the interior monologue of the Creative Subject's True Self.

The Yellow Book

While classical logic was shown inadequate by Brouwer, he did so mainly with reference to the fleeing properites of $\mathcal{U}$ while maintaining the validity of classical logic at least for the case of all events already arrived. That is, Brouwer viewed the "already arrived" as something *finite* for which he maintained the classical principles of logic, including *tertium non datur*. This view point, however, overlooks the real possibility of our *knowledge* of the arrived events itself being a "fleeing property." More precisely, just because something once has been proven to have been constructed, that does not mean that it can be found *now* or even in the immediate future. That is, we may not *now* be able to locate the construction any longer, except by starting all over again, thereby repeating the entire previous existence proof before we may continue. And just because something was done once doesn't imply, analogously, that it can be done again, so one must *prove* the *possibility* of iterating an old proof, especially if the context has changed markedly since the original proof was effected. *Ex eo quod aliqua res est* cognita esse, *non potest evidenter inferri quod alia res sit*.

. . . . .

We say that P is a *"fleeing property"* if

i. For each natural number n, there is a method $\mathcal{M}_p$ deciding between P(n) or its absurdity (not-P(n));

ii. It is not known whether a method $\mathcal{M}_p$ can be written down by means of which a natural number n can be calculated such that P(n);

iii. There is no proof of the absurdity of P(n) i.e. of not-P(n).

By a *critical number* K_p of the fleeing property P we understand the (hypothetically) smallest natural number possessing P. If $n < K_p$, n is called a *down-number* and if $n > K_p$ an *up-number*.

But now it becomes obvious that the property of being the number of all possible proofs itself becomes a *fleeing property*, hence contradicting the finiteness of all possible proofs (as free creations by the Creative Subject herself when considered as sub-spaces C_m of the graph of the species C_n, $n > m$.) Thus, there is a deep contradiction in Brouwer's foundations of mathematics.

. . . . .

As soon as we abandon the use of *restricted quantifiers* many fleeing properties seem to reside in $\mathcal{U}$.

However, soon one makes the experience that by passing to the next universe $\mathcal{U}'$, the fleeing properties of $\mathcal{U}$ are transformed into harmless decidable species. That is, up-numbers of $\mathcal{U}$ together with its critical numbers, become down-numbers in $\mathcal{U}'$.

The species $\mathcal{B}$ of critical numbers of a universe $\mathcal{U}$ forms, in some sense, the *boundary* or *limit* of Σ's knowledge of $\mathcal{U}$ called the *critical boundary* or *limit* of $\mathcal{U}$.

This limit is surpassed by the passage to the immediately next universe $\mathcal{U}'$. Yet, $\mathcal{U}'$ awaits further fleeing properties as soon as restricted quantification is abandoned again.

The length of the sequence $\underset{\sim}{\mathcal{U}}$ of universes is a measure of the failure to learn from the experiences with unrestricted quantification.

The problem is not to be able to extend the sequence $\underset{\sim}{\mathcal{U}}$ indefinitely (that is already taken into attention) but to *classify* the many different ways in which transitions from one universe to another takes place.

Thus we shall expect to deal with a *branching* sequence of universes, each branch modelling a *uniform class* of transitions from one universe to another.

Lecture Notes on Intuitionism (II)

Aspects On The Continuum

Introductory Remarks

We choose an *Initiation Number* **N** as we choose a real line

($\mathbf{R}_x$) $\qquad \qquad \underline{0 \qquad \qquad x \qquad \qquad}$ ...

located around a real point x or with *Weierstrass*' arithmetical notations

($\mathbf{R}_W$) $\qquad \qquad 0,1,2,3,\ldots,\nu,\ldots \infty$

(The fact that the entire line R can be mapped 1 – 1 to the closed interval [0,1] cannot be more astonishing than the fact that all classical and intuitionistic continua are cofinal with an Initiation Number.)

The Yellow Book

We think of the term ν, the *leading term.* as *given* among the terms that are included in the Initial Number N.

Assuming that $\mathbf{R_W}$ is monotone and increasing (if not, drop the superfluous terms) we know that for some term $n < \nu$, $\varphi(n) = \nu$. Letting also φ be monotone and increasing, and such that $\varphi^n(0) = \nu$, φ indicates a *law* or *leading operation* by means of which terms to be included in $\mathbf{R_W}$ are defined. (And the graph of $\varphi^n(x)$ is called the *orbit* of the leading action φ.)

[From a modern point of view one is here dealing with an application of the *number-theoretic axiom of choice* (**AC-NN**)

$$\forall n \exists m\, R(n,m) \Rightarrow \exists f \forall n\, R(n, f(n))$$

The interesting question is, of course, whether f can be choosen as a recursive function or not, i.e. whether or not Church's thesis is an acceptable principle.]

The dots indicated in **R** are explained by pointing to Euclid's first Axiom. Arbitrarily long lines can be drawn and each line is still in the possession of the property of being further extend*able*.

It doesn't follow from the fact that we can draw arbitrarily long geometrical lines, that we also thereby are endowed with the capacity to extend further any one of them. In point of fact, given that we *can* draw arbitrarily long lines why even be interested in *extending* any one of them since a line together with its extension can be obtained as just another arbitrarily long line without any fuss about concepts of "extensions" (and "limits"). (No topology.)

If an arbitrarily given line R can be extended, we must ask, how?

If we assume that the extension obtains by continuing the generating operations of which R is the result, then not only must they be specified (and hence no longer remain completely arbitrary) but so also must the fact that, taken all together, *just* the generating operations suffice for the end result R (screen against objects falling from heaven).

If the original generating operations for R turned out to be insufficient, then *other* means must be sought. In any case, they must be specified and in addition, be proven sufficient (since we are employing more powerful but so far untested means).

In any case, we are moving towards the *topology* of the geometrical line making it come nearer the behavior of points on the real infinite line.

. . . . .

Turning now to the intuitionistic concept of the real line, consider that each initiable *mathematical* activity starts from a fixed *Initiation number*, N (*Anfangszahl*, Hausdorf, 1907).

Once the beginning of a series of Initiation numbers has been permitted, there is no end to the mathematical possibilities so opened up.

Once initiated into a beginning of a mathematical activity, $\mathcal{A}_N$, we shall move in an orderly fashion from one mathematical reality to another, where the latter reality remains just a *possibility* as long as we *remain* in the former reality.

The jump from one mathematical reality to another is obtained by exhausting a reality which at present is at hand until a limit is reached and from which another reality is formed.

The search for mathematical possibilities is moved by a desire to expand mathematical reality. Every new-formed mathematical reality contains the key to any further mathematical possibilities. It is the possibility of initiating a mathematical activity again and again which our understanding of the (increasingly) varying mathematical reality must be based on. In particular, it is the reality of the form*ed* limits in which the mathematical possibility operators reside.

Each process realized by a mathematical activity $\mathcal{A}_N$ is *cofinal* in N (or one of its initial (singular) segments).

In particular, if $\mathcal{A}_N$ realizes a continuum, C, the process C is cofinal in N.

The initiation number, N, of a mathematical activity is usually a limit point of a previously introduced activity $\mathcal{A}_{N'}$, based on an initiation number N' which was obtained as a limit of a still earlier mathematical activity.

If N' forms a *proper initial segment* of N i.e., if

$$N' \longrightarrow < N$$

and is not cofinal in N then $\mathcal{A}_N$ is said to be a *regular* mathematical activity and following upon N' in the ordering of initiation numbers $< \eta, \longrightarrow < >$.

If N is an initiation number which is not cofinal with an initial segment of any previously obtained initiation number, then $\mathcal{A}_N$ is said to be *singular* mathematical activity.

The simplest possibility of a singular mathematical activity obtains with $\mathcal{A}_N$ as a limit of an unending but well-ordered chain of regular mathematical activities with regular initiation numbers N_{00}, N_{01}, N_{10}, ... such that

The Yellow Book

$$N = \varprojlim N_{00}, N_{01}, N_{10}, \ldots$$

and N is cofinal with N_{00}.

Remark. Without the Axiom of Choice, there are (classically) no regular limit numbers above ω ($= N_A$) and only 0 and 1 below. Thus, "2" (= "10") is the "smallest" singular ordinal number.
. . . . .

The first singular initiation number $N = N^0$ is the *first internal* limit imposed on the seemingly unending chain of increasing mathematical possibilities which all begin as regular mathematical realities.

In order to transpose a possibility of a manifold of mathematical realities into a manifold of new mathematical possibilities we must be initiated into a mathematical activity which will contain the latter manifold as a reality.

None of the previous regular activities contains the new manifold as a reality but at the limits of the said regular activities we shall construct a "diagonal" thereby delineating the internal limit as well as the construction of the first singular initiation number N^0.

Because of the breakdown of the monotone increasing cofinalities of the initiation numbers $N_{00}, N_{01}, N_{10}, \ldots$, the splendid regularity of the formation of new mathematical possibilities has been spoiled and we are about to start contemplating an entirely new form of substitution so that a sequence of regular initiation numbers, all larger than any of the previously obtained initiation numbers, can be picked up again.

For a traditional mathematician, of course, there is only one concept that comes to mind: Cantor's classical concept of *Diagonalization*.

But, then, the first question is: how do I know that the diagonal construction is *unique* and how do I tell apart two such constructions if they are not equivalent?

Yet, the diagonal construction exhausts the possibilities contained in the initiation numbers $N_{00}, N_{01}, N_{10}, \ldots$. It is this power of N^0 which defines the reality of a first internal limit of mathematical possibility.

All the realities that have been encountered by following the sequence $\mathcal{A}_{N_{00}}, \mathcal{A}_{N_{01}}, \mathcal{A}_{N_{10}}, \ldots$ have been *regular* realities. Nothing upset the regularity of mathematical reality until someone delimited the work so far layed down by laying out a diagonal following the exact contours of the limits of regular reality.

If you are not permitted to transgress the diagonal contour of the work you have so far layed down, then, indeed, you have exhausted all your possibilities and reached the limit of the realities encountered so far.

Yet, it seems, by staying where you are, i.e. on the diagonal, mathematical reality is not becoming enriched. Rather, it stays fixed and you have taken a step out of it, seeing just the far end contour, as it were, from a distance.

In order to find yourself back in mathematical reality again, you need to be initiated into a mathematical activity which has a replica of this contour well within its limits and while rising towards those limits you may survey the totality of your previous mathematical reality up to and a bit beyond the fixed diagonal contour.

Since there is no possible communication between these two mathematical realities, i.e. the latter contains an initial assumption which was not permitted by the reality of the diagonal world, we must be initiated into a *third* mathematical activity in terms of which all communication between the diagonal world and its "successor" takes place, i.e. a mathematical reality which treats the reality beyond the diagonal world as a regular *extension* of it. But this is where all the trouble comes in.

When I speak of picking up again a regular sequence of initiation numbers, I speak of a mathematical possibility which is formed out of a reality which constitutes a conceptual break with any mathematical reality encountered so far through the sequence $N_{00}, N_{01}, N_{10}, \ldots\ldots$

Lecture Notes on Intuitionism (III)

According to Brouwer, intuitionistic mathematics must be regarded as a *free creation* by the human mind.

The freedom with which the Creative Subject constructs mathematical objects is only paralleled by her freedom of choosing the *rules* upon which the existence of her constructions depends. Although freely chosen, these rules need not correspond to any given mathematical object although the result of following the rules may do so. It is even possible to consider freely chosen rules the following of which certainly does *not* correspond to a construction of a mathematical object.

It is not quite clear how free one is in making a distinction between the latter kind of rules and those the following of which leads to a mathematical construction. It is indeed possible to consider cases where one may consider the result of following some rules under both aspects, i.e. the "same" construction κ is connected with two distinct objects, say, a text $T = \{s\}$ and a natural number $n = s$.

In such a case the Creative Subject is not completely free to interpret the resulting construction because unrestricted freedom of interpretation would introduce a possibility of deep confusions.

Thus, because of the requirement of the *clarity* of intuitionistic mathematics, the latter cannot be regarded as a *languageless activity* as stated by Brouwer. This does not mean that, say, the intuitionistic continuum, C, will not eternally remain without an exhaustive description $C[\tau]$, but only that cases of ambiguity are resolved by introducing specific rules for the use of certain combinations of signs.

It follows from the above that the logical approach to problems in the foundations of intuitionistic mathematics is by *regressive logical methods*. That is, the Creative Subject *first* arrives at an intuition of the continuum C and only *thereafter* does she arrive at the intuition of a description $C[\tau]$ of C. While the latter description necessarily remains incomplete, nevertheless, in the course of constructing a description $C[\tau]$, the Creative Subject encounters a description b^τ of objects B which are finite substructures of C.

One unambiguous such object is the *empty continuum*, Λ_C, described by a fulfillable intuitionistic method, $\mathcal{M}_{\Lambda_C}$, corresponding to the *first act of intuitionism*, while another *ambiguous* such object would be the process of events defining the series of *initial numerical signs* known by "heart."

(K_h) 1,2,3,4,5,.....

That is, does a description b^τ of the series K_h correspond to a finite mathematical object or must the "object" K_h be identified with some species of "pre-ontological" objects? That is to say, the problem is whether or not the Creative Subject prefers to take the initial acts out of intuitionistic mathematics proper. If she does, then b^τ does not correspond to any mathematical object and the symbol K_h becomes mathematically unambiguous in the clear sense that *it* does *not* belong to the language of intuitionistic mathematics. But this must also mean that the Creative Subject has lost her freedom to rely on the initial acts of intuitionism when effectively considering the existence of mathematical constructions. That is, the Creative Subject is *forbidden* to consider an intuitive Nn such as K_h.

Again, the existence of this possibility shows that the complete freedom with which the Creative Subject constructs her mathematical objects is fictitious and, therefore, stands in need of an adequate formalization. In particular, the logical mysticism surrounding her initiation into the abstract activity of the beginning steps of intuitionistic mathematics stands in need of a detailed formalization.

Appendix II

MĀ-Theater

Theater of the Eternal Mind

MĀ-THEATER:
From Abstract Nō To Abstract Butō (1968–1988)
[ABENDSCHAUSPIELE/COGNITIO VESPERTINA]

According to Zeami, dance is instructed by song, i.e. poetry. Thus, if dance is instructed by *abstract symbols* instead of (concrete) words, *abstract dance* is created.

Λ is a symbol for the Empty Dance

(Utsu Butō - Void Butō)

The stage on which the Creative Subject shall be born is a lattice of (existential) contradictions—a vacuum encased by darkness.

Abstract Nō begins with a study of Non-Being and ends by a presentation of the character **MA**—annotated with its appropriate X-index, $M\bar{A}_X = M\bar{A}$. More generally, "**MĀ**-Theater" stands for "**X**-indexed **MĀ**-*Theater*" of which Abstract Nō is a *separable sub-species*.

By contrast, the X-topocosm of *Abstract Butō* is characterized by (the consequences of) the singular identity

which is also written

$\bigcirc \equiv \bullet$

Remark on *Ankoku Butō*. The character **ANG** signifies, with complete silence, what the concept "Day" stands for—while the character complex **ANKOKU** signifies the darkness brought about as the sounds of civilization already at dawn pierce through the daylight of what was once true.

Ankoku Butō-no-Zhuang Zhou from the Flowers of the Black Chrysanthemum (1985) is my second composition in the style of *Saido-Butō* and it too is characterized by the abstract identity

$\bullet \equiv \bigcirc$

where " $\bigcirc$ " now also carries the connotation of "Day."

The Yellow Book

Circles of Being: ○ ○ ○ ○ ○

Circles of Darkness: ● ● ● ● ●

The Circles of Being and Darkness are unified in Abstract Nō by the identity

$$間 \cup 有 = 闇$$

Reminder on Yugen. Yugen applies to speech, to the sounds of words. Yugen signifies the deep beauty of the darkness in which the sound of words envelope the Day.

Being, shattered by the sound of words — the flowers of a Black Chrysanthemum.

Saido-Butō designates the continuous trajectories of the infinitesimal arrow arabesques of a shattered Being falling through the infinite darkness of the mind's dislocated centers.

(Remark on the formal derivation of the "character" **SAIDO-BUTŌ**. The use of the character **SAIDO** within the theory of Noh goes back all the way to Zeami by whom it was used to signify the concept of a *"pulverizing movement"*. Resorting to a compact and concise notation

$$有 = 砕動 \quad \text{(SAIDO-BUTŌ)}$$

is thus the general formula by which an ancient dance step of utter darkness becomes a "pulverizing movement" as indicated by Zeami's character **SAIDO**.)

Arrows of movements for a vertical topocosm
(The Static Anfractous Flows Of A Cosmic Current.)

These are the principal arrows of movements in the style of Saido-Butō in which emphasis is on very small but widely differentiated *vertical movements* which are subjected to an equally small *anfractous rotation*

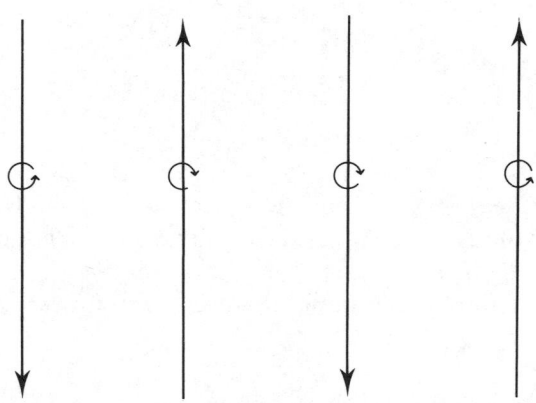

Movements in this space of vertical rotational movement gradually become displaced as instances of *locally infinite spaces* encased by an inpenetrable darkness into which the Creative Subject eventually disappears. The Creative Subject creates absence by being present only within the confines of a locally infinite space in which each movement pulverizes its very presence. More specifically, the pair of arrows which determine the vertical rotational movement is composed in such a way that vertical and horizontal movements alternatively absorb each other. That is, given an indexed family of a principal arrow, say,

the Creative Subject recurs and recurs as a limit point

•

of motion and stasis by which is defined the utter darkness of the enveloping space of the X-indexed family of principal arrows

Historical Reminder. (Pre-socratic Dance.) The determination of movement and stasis by poetry goes back at least to Vedic times. It is in the Vedic texts that one first encounters a refined concept of *correct action*, i.e. **ṛta**. The Vedic Dance remains largely unknown as do the later Eleatic Dance or the Eleusinian Dance. However, it is known that a theme common to all these ancient arts of dance is the action of transformation by sacrifice. Thus, it must be expected that *correct action* (**ṛta**) translates for the Creative Subject as *total personal transformation* in a rigid performance of Abstract Nō or Abstract Butō.

常住不滅

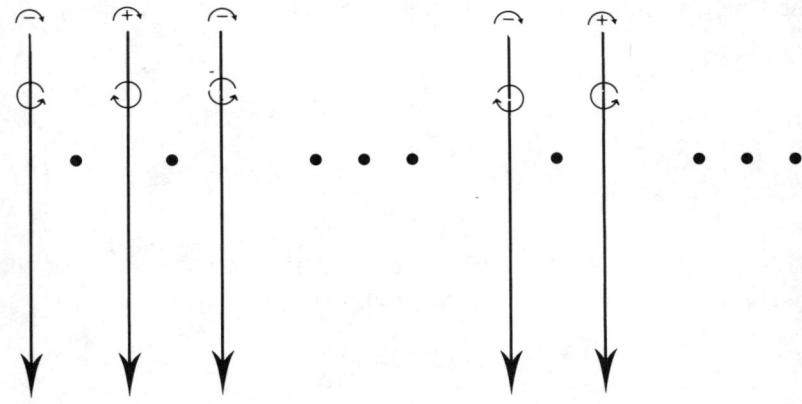

(Elements of Movement $\bar{X}$).

Ankoku Butoh—no—Soshi
(Tatsumi Hijikata in Memoriam)

>You do not dip twice in the river
>Beneath the same tree's shadow
>Without bonds in some other life.
>
>   Motokiyo

Characters:

Zhuang Zhou reincarnated as dream-images	**Shite (1)/Shite 2**
Dharmakirti	**Waki**
Wanderer (Sophist?)	**Tsure**

Space:

The interior of a recurrence of *Zhuang Zhou's* eternal dream

Season:

Autumn

Chorus: November

Steps

Abraxas

In the

Fall ...

The last task

To ask the

Questions never

Thought, appears what

Remains in these

Ruins of human hopes &

Fears

in the

Fall ...

        (Among the evening's

            Echoes.)

Dreams of the eternal
Maps have all ceased.

Words are gone.

Lacking an end they also lack a
Beginning. Only the broken

Thought of a middle without

Beginning ... nor end remains as

a single

Possible dream of eternal

Thoughts without words that

Fall ...

Abrasax !

Abraxas !

          ....

Two   (**Tsure** enters)

Utter

Ancient

Emptiness

Memories

Time

Night ...

Night ...

Two (**Waki** enters)

Memories

Ancient

Emptiness Emptiness

Memories

Time

Utter

Time

Ancient

Two

Utter

Night ...

Two shadows passing each other

Utter darkness contained in   in the

                                                             a single

Ancient step that continues

                                                             forever in the

Emptiness of space divested

                                                             of any

Memories of the passage

Time during the passage   of

                                                             of an endless

Night ...

Waki: No one rides before, no one comes behind

and the path bears no fresh prints

How now, am I alone? Ah yes, I see:

the path which the ancients opened up by now is overgrown

and the other, the broad and easy road, I've surely left

Yet, as I crossed this   (Shite enters)

bridge of many dreams,

I realized - for the first time -

That even for a dream, that is, any

Dream, to be real, **it** -
Reality! - must also be a
Dream. Because, only what is
Dreamed is eternally
Real ...

Chorus: Every dream is a dream, although every
Reality is not real
because a
Reality not dreamed
is not
Real

C/W: Indeed, certainly every
undreamed reality
remains
Unreal ...

Waki: Yet, notable transgressions exist, such as the ominous moment when a
Dream turns into a
Nightmare - which is a dreamed
Reality consisting of

Memories of something
Unreal, of something not yet real,
Of something with a beginning
& a middle but with no

End.

Aprameya-vasunam

Aviparita-dṛṣṭih

W/C: **Vikalpa-yonayaḥ**

Śabdā vikalpāḥ

Śabdā-yonayaḥ

Chorus: A

> Dream
>
> Nightmare
>
> Reality
>
> Memories
>
> Unreal
>
> Of
> &
> End.

Tsure: Sarvam mithyā bravimi.

> (Among the evening's
> Echoes.)
>
> I believe I see the
> Image of **Master Zhuang Zhou** - this time
> Reincarnated as an
> appearance of an eternal
> Reality!

Waki: Ah yes, I now believe

> I see the eternal image of **Master Zhuang Zhou,**
> Too! - Maybe he can

Guide my way along

This untrodden path of a

Dream which perhaps is

No longer real ...

Chorus: Yes

I

Too

Guide

This

Dream

No

Shite: Among my memories, my initial

Memory was of a

Dream in which I entered, effortlessly,

Immediate Reality - without knowing

Anything about what is

Real - except my dreams - except my

        Dreams of

        Reality ...

Chorus: Every dream is

                                                    real.

            Every dream is a dream.

                                                      Like

            Mirrors of the eternal maps

                                                    of recollections
                                                    of

            A receding reality

                                                    of ancient mirrors ..
                                                   of ancient dreams ...

Shite: Yet, how ominous, but I cannot

            Recall anymore **if** I

            At that moment of my noble birth,

            Was truly dreaming about reality,

            That is,   **this**

            Reality,

Or,

were I actually in a state of

Dreaming about **SOMETHING ELSE?!** – of which

S/C: Reality cannot know, or, perhaps, even,

Must not know?

Shite: **Or,** again, is this all an

Unsurpassable never ending

Nightmare – such that, if I,

Zhuangzi, am not a

Dream, then my being is

All the same a

Dream about a non-dream, a

Dream about an ominous **Butterfly** which cannot

Be a dream by itself?!

. . . .

Shite 2: There is no Butterfly as there is
no (-) Self!!

But, there is , there remains , as the

Creative Subject's shadow , a single

Endless , solitary , ancient step

S2/C: That continues on and on in

Utter darkness until the step itself is

Dissolved by its own impenetrable solitary

Darkness as if the

S2/C/T: Dream was not real . . . .

. . . . . .

(The two Shite characters begin their dance.)

Chorus: Dreams dreamed - dreams of dreams - are   not always

Dreamed dreams - not always, not always -

Whether or not the passage

        is crossed, the

Passage between what is no

        longer

& that which is still not yet,

        the

Crossing of the bridge

        of

Dreams ...

There

                                                  there

shadow

ancient

                                                  in

itself

Dissolved

                                                  as

real ...

Two shadows again

                                                passing

                                                each other

                                                in the

Utter darkness contained

                                                in a

                                                single

Ancient step that continues

  forever

                                                    in the

Emptiness of

                                                  space

                                                divested

                                                of any

Memories of

                                                the

                                                passage of

Time during the

                                                passage of

                                                an endless

Night ...

THE END.

A Non-unique Extraordinary Set
(From Toposes and Adjoints, 1976)

Notes Toward a Translation of Parmenides

Charles Stein

Fragment 1

The mares that are able to take me
                as far as I want to travel
had so taken me
once they'd set me down on the Daimon's Way —
for it is *She* that takes the Knower through each town.
Onto such a route had they placed me
and the knowing horses carried me along it, straining
                                    at the reins.
And the daughters of the sun went before us, leading the
                                            way.

The axle of the chariot
    urged round by eddying wheels attached at the ends
                                    put it in motion
and the axle whistled and shimmered as it turned in the
                                              nave
while the daughters of the sun sent us into the light
        having come out of Night's abodes
and pushed back the veils from their faces with their hands.

Up *there* are the gates of the tracks of day and night
fitted above with a lintel
                and below with a threshold of stone
and the openings themselves, high up in the air,
            are closed by mighty doors.

Dike—The Equalizer—holds the keys to them.
And the sun's cunning daughters
        used mild speech to persuade her
                to open the gates.

The gates, when opened, opened on a vast expanse
and the daughters of the sun
        drove the chariot and mares out on to it
and the gates were fixed on singing axle hinges.

And taking me by the right hand she spoke to me thus:

"Oh Youth, linked with your mares to immortal charioteers
who have lead you here to my home—Welcome.
Since it is by no means an inappropriate destiny
        that has sent you forth to travel this path
                far from the wanderings of mortals
        but a Right and Just one,
it is necessary for you to learn all things—
both the stable heart of well-rounded truth
as well as the notions of mortals—
   (and in these there is nothing at all to put your faith in)
nonetheless you shall study such matters also—
how the *things that seem*
        (and these pervade everything)
                must seem *to be.*

Fragment 2

But come, and I will instruct you.
And you must take back home with you what I say—
whether in fact there are only these two ways for thought
                              to travel:
Either:
    "It *is*"
        (and also) "That it *is* not"
                      is not

(and this is the path of which one
        ought to be persuaded
    for it leads to truth)

Notes Toward a Translation of Parmenides

Or:
      "It is *not*"
            (and also)
                  "That it *is*"
                          cannot be

(and I say that this is an unconvincing road:
                    it doesn't turn.)

You cannot know what in fact has no being—
        this is impossible
               and you cannot speak if it.

Fragment 3

For "to know" and "to be" is "the same."

Fragment 4

Consider things, which, though being far away
    are nonetheless certainly present to the mind.
For you shall not cut off being from its continuity
                        with itself.
It will neither be dispersed from, nor contracted within
                  its kosmos.

Fragment 5

It is all the same to me where I begin
      for there shall I return.

Fragment 6

It is necessary to know and to say that Being is
for the other way—the thought that it is not—
                          cannot be.
It is.   And "cannot be"   *is* not.

Think about this.

And now I must dissuade you
      not only from taking *that* one of these two courses
but also from another
      upon which mortals wander double-minded
                          not knowing anything
for only ineptitude straightens
                    the errant thought in their minds.
They are carried along
            blind and obtuse—
                    these utterly astonished ones
                    this indiscriminate hord—
by whom the "to be" as well as the "not to be"
are thought of as the same
and
     at the same time
                not the same

and that the track   down which everything passes
                        is backward turning.

Fragment 7

For you are never going to put *this* thought to rest:
    that things that are not   are.
So urge your thoughts off this duplicit pathway
and don't let well-worn habit force you down it—

don't let your eye wander don't let your tongue wag
        don't allow your ear to echo aimlessly.

Rather judge by thought
the much disputed proof which I have spoken.

Fragment 8

There is only one path left
and that is
that   "it is."

Notes Toward a Translation of Parmenides

And on it there are many indications
that Being is unengendered
that it cannot be broken apart
        (for it is whole, without parts)
that Being does not fluctuate,
that it has no end.

It never *was*.   It never *will* be.   It is all NOW—
                                    one continuum.

For what kind of engendering can be sought for it?
How and from what source might it have grown?

And do not say and do not think
        that it came out of not-being
for it cannot be said and it cannot be thought

        that "it is"   is not.

And what need could have driven it to grow
        starting from nothing
           at some time earlier
                rather than at some
                        later time?

. . .

[[What debt incurred, stirred it to grow

then, rather than at some other time

if it first came into being out of nothing?]]

. . . .

Therefore: either "it is"
                or else "it is not"
        and in either case
                completely so.

Nor shall strength of trust
        incite anything   to come into being
           out of   not-being
                severed from "it is" itself.

On account of this, therefore, Dike neither loosens her
                                                              grip
      so that Being might come into being
   nor does she loosen it that Being might perish
                                but she holds fast.

And the crisis in these matters lies in this:
        either "it is" or "it is not."

Now surely it has been decided
            according to what is necessary
  to abandon the nameless and unthinkable alternative
   for a true way   *that* is not
            and that the other way: that "it is"
         is real and genuine.

How could what is
            afterward pass out of being?

      How might it come to be?

For if it *came* into being,   then it *is*   not.

And it is not also
        if it only is *going* to be.

So the alternative "coming into being" is extinguished
and of "going out of being"
                       nothing can be discerned.

. . .

[[Thus growth is extinguished
     and destruction
            stopped.]]

. . .

Nor can Being be segmented
since it is altogether, of a single piece.

Nor does it excede itself   or fail to reach itself
      thereby disrupting the continuity of itself.

Notes Toward a Translation of Parmenides

But Being is everywhere replete, everywhere continuous
and all of it is connected
                close
                      to all of itself.

But without motion
                locked within the limit of mighty bonds
        anarchic   without beginning
                and not to be brought to an end
(since both "coming into being" and "passing out of being"
                        have been banished far away by
                                    true belief)

the same in the same   remaining with itself
                      it lies where it is

for strong necessity has locked it in limiting bonds
        and imprisons it all about.

Nor is it lawful for Being to remain uncompleted:
                Being lacks nothing.
For if it lacked one thing it would lack it all.

Because of this, therefore, to think and the thing thought
            are the same thing (namely Being)
for you will not find thought without the being about which
                that thought is uttered.

And nothing is or ever will be outside of what is
since Fate has fastened Being to remain
                            a motionless whole.

And all the other designations
                which mortals have laid down
        having trusted them to be true
            are only names

namely: "coming into being" and "going out of being"
        and   the mixture of the two   not-being and Being
                                          together
            alteration of brightness and color
                and change of place.

However, since there *is* an uttermost limit,
"It Is" is fully established on every side
resembling in dignity and mass a well-rounded sphere
equally distributed, balanced, in every direction
for it must not come out somewhat greater in one place
      somewhat smaller or more humble in the
                  next
as neither does not-being exist
     to prevent it from accosting its like
nor can Being be either more or less than Being is
    in one place as over against another place
  since all of it is inviolable, equidistant from
           everywhere
  and all of it alike within its bounds.

At this point I stop such thought and speech
     as you ought to put your trust in
concerning truth. From here on, learn the beliefs of mortals
listening to the deceitful ordering of my words.

For they set up two forms in their minds for the business of
              naming
whereas even one form is incorrect. In doing so they have
      strayed into errancy.
They discriminate bodily antitheses and set up signs
apart from one another: on this side the bright tongue of
             flame
being gentle and most light, in all ways the same with
            itself
and in no way the same with another; and on the other side
          in opposition
thick night. Solid. Massive. Bodily.

I speak to you here of what is merely a plausible cosmos
so that no thought of a mortal
      may ever get the best of you.

Notes Toward a Translation of Parmenides

Fragment 9

But since all things have been dubbed "Light" and "Night"
and according to their powers, this pair has been
                          attributed to various entities
all is full at once of light and nocturnal obscurity—
full that is of both—since nothingness
                                  nothing
                                           has a share in.

Fragment 10

You shall know of the aetherial nature
                and of all the signs in the aether —
and of the inapparent acts of the pure torch of the
                                         spotless sun
                and from what it came into being.
You shall learn of the wanderings
                      of the round-eyed moon and its nature.
You'll know from whence the encircling heavens grew
           and also how Anangke, guiding it, bound it
                  to hold the limits of the stars.

Fragment 11

How Gaia and Helios and Selene, Aether and the Galaxy,
outermost Olympos and the hot strength of the stars
                                  rushed into being

Fragment 12

For the narrow rings are filled with unmixed fire
and the next ones are full of the night
                and a tongue of flame shoots out.

Charles Stein

In the middle of this is the Daimonness
      who steers all things
and she commands the commingling
    and the hateful births of everything
   sending the female to mix with the male
      and then contrariwise, the male with the
               female.

Fragment 13

She devised
    Eros—absolutely the first of all the gods

Fragment 14

Night-shiner—
     about the earth
        wandering
          —an alien light

Fragment 15

always straining after the rays of the sun

Fragment 15a

water-rooter

Fragment 16

According to the disposition of bodily parts—the limbs
        that wander everywhere—
is the mind present in persons.
For the nature of the limbs which thinks in persons
     is the same in all and each.
And the full is thought.

Notes Toward a Translation of Parmenides

Fragment 17

On the right sides—males
        on the left sides—females
                (sides, that is, of the womb

Fragment 18

(Latin fragment)

Fragment 19

Thus according to opinion, these things gestated
                        and now exist
and hereafter they shall grow and later perish.

And for each of them humans have established a distinct
                                                  name.